Surviving the Age of Virtual Reality

Surviving the Age of Virtual Reality

Thomas Langan

University of Missouri Press
Columbia and London

Library of Congress Cataloging-in-Publication Data
Langan, Thomas.
    Surviving the age of virtual reality / Thomas Langan.
        p. cm.
    Includes index.
    ISBN 0-8262-1252-2 (alk. paper)
    1. Meaning (Philosophy) 2. Civilization, Modern—20th century.
I. Title.
B105.M4 L355 2000
909.82—dc21

                                                                99-054265

∞ ™This paper meets the requirements of the
American National Standard for Permanence of Paper for Printed
Library Materials, Z39.48, 1984.

Designer: Vickie Kersey DuBois
Typesetter: Crane Composition, Inc.
Printer and Binder: Edwards Brothers, Inc.
Typefaces: Serpentine Bold, Serpentine Sans, Amasis MT

To Bruce A. Stewart, KtB,
whose insights, arguments, and expressions are woven
through this work, in Friendship and Gratitude

Contents

Surviving the Age of Virtual Reality

PART ONE
Where Are We?

Chapter 1

Introduction
Welcome to the HTX!

You Are in the HTX and Did Not Even Know It!

As we get ever more swept up into the brave new worldwide society, even some techies are concerned that it may be whirling out of control. (And they started to feel that way before discovery of that millennial warning from God, the Y2K bug!) This world system is so new and powerful—growing exponentially as the population explodes and interconnectedness expands—and so unprecedented, and hence unpredictable, we do not even have a name for this new kind of planetary social phenomenon.

Or we didn't, until now: Welcome to the HTX! This new social phenomenon has roots as old as man's first technical inventions; no one can establish a border for it, there is no moment when modern industrial society became the HTX. But our research group of techies (philosophers and even a psychiatrist) after struggling for several years to describe it, drifted one night, under the influence of low-tech Chardonnay, into calling it the HTX: HT for High Technology and X, meaning "No Name." And, we realized, it "X's out" cultures and civilizations. It roots no one in the soil, rather it obfuscates our natural roots, hence it is not a *cultura*. And while it spreads its net out from many urban nodes, it is no civilization, for it is not now bound to a city, or even cities as such, indeed, again unlike civilizations, it has no center. It is penetrating old cultures and transforming great civilizations, especially the modern European from which it issued, interacting with them in ways that reverberate back on, and thus alter, the HTX itself (the Japanese are forced to be more American and Americans more Japanese as the HTX transmits profound influences from both). Now it is even being accused of forming an entire generation, "Generation X"!

It is not a culture, not a civilization, so *what* is it then?

Answering that question is actually vital to your personal prosperity, for the HTX, whether you like it or not, forms the context in which you live (and invest). A proper knowledge of it is necessary for your survival (and prosperity) and crucial to the maintenance of a civilized society of the kind you might want your children to live in. (Civility may belong to a waning civilization, just as Western forms of courtesy stem from an earlier, more Christian phase, but if we do not fancy nihilistic brutality, we need to ask how can we "civilize" the beast.) Upon our *wisdom* in understanding and our *prudence* in managing ourselves within the HTX and steering of the HTX itself in acceptable directions depends the very survival of mankind. None of us wants what is left of Western European civilization HTXed into forms we cannot even bear to imagine: for instance, a future in which children are the fruit of designer genes and everyone is so busy managing there is no time for family or serious friendship. Nor do we relish the thought of the power concentrations we have achieved being unleashed on missions of unimaginable destruction.

Oversimplification is always dangerous, and in this case, it is especially foolish. The twenty-plus essential dimensions of this X we were able to identify (which I'll get to later) are interacting, weaving a dense, dynamic web, the vectors of which processes we need to grasp, and see better their interactions. But already most of the best heads, from CEOs to university types, are so dis-tracted (literally: "pulled apart") running after their dispersed responsibilities, they have neither time nor appetite for too heavy an inquiry into this reality molding their lives.[1] So, facing reality, and since I want the best heads to read this study, I shall keep it short, and the gobbledygook minimal, foregoing many of the interesting side issues. But, unavoidably, the question is heavy, because it is about *Being*. (Forget that "lightness of being" nonsense! Reality is complex and demanding.)

"HTX" is the code name for the *Being* of the epoch in which we are trying to survive. This era has become planetary because a *world system of*

　—large *institutions,* with

　—rich *traditions* behind them, are interacting on a planetary scale, and as a result of efforts to express themselves, evolving

　—*symbol systems*—from common languages (especially "bad

---

1. Two books offering insights into how things got this way are Walter Truett Anderson, *Reality Isn't What It Used to Be* (San Francisco: Harper, 1990); John Ralston Saul, *Voltaire's Bastards: The Dictatorship of Reason in the West* (Toronto: Penguin, 1993). The first deals with the inability to work out what Eric Voegelin calls "a new civic theology" for the West, a common core of beliefs that makes some defensible sense out of human existence, forming a point of adherence about community matters. Without it, how are we to revitalize the locomotive that pulls a coherent society along, projecting it beyond immediate gratification, which is just practical nihilism? The second explores why managers are the way they are.

English") and liturgies to mathematical systems and computer languages used all over the world—and

—*managerial tools* for trying to identify and control
—worldwide *processes of change.*[2]
—All of this bathes in a *new light* of being, new
—*mind-sets* brought to the scene through acts of billions of individual human agents looking at the world through similar interpretative horizons, and affecting one another through living in a shared planetary context.

Obviously, then, the HTX is not a thing, like a tree, nor is it simply a set of related elements assembled to form a system, like a telephone network. (It does contain many such systems, and accounts for their existing and their form, including even the "world system" of interacting international institutions.) It is more ethereal than power lines, more like the friendship that binds two people—ethereal, constantly evolving, but not just a "worldview," for it has an enduring, striven-for core. Although more spiritual than superhighways, even "information superhighways," it is very *real.*

The nature of that "reality" is the issue. It is in that reality we must make our way. Coping with life and dealing with reality—indeed learning *to see* and *getting reality right*—are much the same. I put the now-tired phrase "virtual reality" in the title as a symbol of the HTX's power to create new kinds of reality. So the question of the HTX's own reality holds the key to how we can—and should—relate to it. (This book will face ethical questions of how we ought to live in the HTX if we want to attain certain ends. I shall argue for the goodness of certain ends, explaining why I believe they are *really* good. They follow from certain conclusions about reality itself, and endeavor to be re-sponse-able. What gives this ability to re-spond? And what constitutes an adequate re-sponse? Would you believe the Latin root *spondeo* means "I commit"? We cannot see the HTX without engaging in it, and at the same time withdrawing from it enough to make it our ob-ject, "thrown over against (*ob*)" our critical regard.)

The reality of the HTX might be described as "spiritual" in the way the monetary system is spiritual: it exists primarily, not just as computer entries, but in the mind! Think of what is being opened up and how old attitudes are affected by just one HTX hardware dimension,

---

2. A process is a course of change long lasting and consistent enough to call itself to our attention. See my *Tradition and Authenticity in the Search for Ecumenic Wisdom* (Columbia: University of Missouri Press, 1992), 19–20, 54, 92, 135–39

call it the "optical": the new powers of vision introduced by devices such as the electron microscope, film, the computer, television, and MRI machines. Through them, our perceptions, our very ability to see, remember, forecast, and represent are being expanded. Even in producing a virtual reality like a flight simulator, one is still creating an imaginary space, a nonlinear narrative.

But HTX concepts and attitudes, spiritual though they be, do end up affecting the physical landscape: the cities, road networks, factories, and the power lines are results of HTX thinking made tangible, staked out in cosmic time-space, and directing energies. Today we even have to create new cultural objects, wilderness preserves, to retain patches of nature not totally transformed by the HTX, but even the wildest of them is "managed." (Management is a mind-set, a way of looking at, and transforming, reality.)

This book will be useless unless it evaluates this HTX milieu in the light of a wisdom more fundamental than what you usually find in pop studies, even one as insightful as Neil Postman's wonderful *Technopoly: The Surrender of Culture to Technology.*[3] I am convinced a wishy-washy developmental humanism like Postman's will not save us from the bad fate, the dehumanization, he describes in convincing and scary terms.

A recent article in the *New Republic* offers this warning:

> We are living at a time when all the once regnant world systems that have sustained (also distorted) Western intellectual life, from theologies to ideologies, are taken to be in severe collapse. This leads to a mood of skepticism, an agnosticism of judgment, sometimes *a world-weary nihilism* in which even the most conventional minds begin to question both distinctions of value and the value of distinctions [emphasis mine].[4]

I shall not waste time impressing upon you the dangers of such a nihilism, of which mindless terrorism is just one of the most obvious effects. And it would be silly to blame high technology too much for the negative in what is happening. It is a grave distortion of reality to underplay the accomplishments, not just of HT but of all of modernity, including "liberal society." Here in "postmodernity" is where our grandchildren are going to live (whether liberal society will survive for long is another question), so we must struggle to find a solid founda-

3. Neil Postman, *Technopoly: The Surrender of Culture to Technology* (New York: Vintage, 1993) contains many good insights and should not be overlooked.
4. *New Republic* (Feb. 18, 1991): 42. Cited by Postman, *Technopoly.*

tion within it for building and leading meaningful lives, not just personally, but societally, for meaningful life will prove impossible in isolation. If we do not find a defensible meaning to life, we cannot educate, but only technically train our children. Then we shall not bear to live with them when we are old. Well-trained, technically adept children would probably solve that generational problem by euthanizing us.

Culture was rooted in the land and that which derives from land (think of the patronage of Mozart and Goethe, for example, even that late into civilization). Civilization, on the other hand, is rooted in the city (dependent on, and also affecting underlying cultures.) At the same time, civilization de-grounds experience of more basic process, as it produces "high culture." (We shall explore in the next chapter the sense in which cultures become "higher.") Finally, the HTX is warped along the network—it is "an overlay structure," drawing on, penetrating into, and transforming the civilizations and cultures that it overlays, including its mother civilization, the occidental.

Table 1

THE DIALECTIC OF VERTICAL TRANSCENDENCE

NATURE

HUMAN NATURE

CULTURE

HIGH CULTURE AND CIVILIZATION

INBREAKING OF DIVINE TRANSCENDENCE FROM BELOW AND FROM

ABOVE

HTX

The HTX is a planetary "world" embracing smaller worlds. It is the form of what we mean when we speak of "THE world." There are many kinds of little worlds, ranging from the individual's (as in "he is lost in his own little world"), through small rule-bounded worlds like that of a baseball game, and the more open, less explicit rule-bound world of a family, and larger worlds, like that of American federal politics, or the world of information technology, to vast planetary worlds with ancient roots, like the world of Catholicism or the HTX itself—THE WORLD.

Each kind of world has its own kind of being, which, in all cases, envelops the basic being of human nature. It is only individual human beings who project worlds and make them go, and groups of human beings who make all the intersubjective worlds possible.

I am obviously invoking here a special sense of "being," meaning something like the "light" that illumines a given "world," supplying its context and meaning. Be patient! You shall see what I mean in a moment. For now, just think how the intersubjective light illumines a baseball game, brought to it by players, umpires, and fans alike, cooperatively, and you will see that that particular light is quite different from the religious light of a Eucharistic celebration, when the intention to invoke a special kind of divine presence puts one in a sacred world, and how neither is reducible to the mere things involved, diamond versus sanctuary, uniforms versus vestments, balls versus hosts. And an umpire plays a quite different role than a priest. Umpires who get too pontifical are not appreciated.

Yet no world is ever an island; it relates to bigger worlds, both through the people who incarnate them, who live in those other worlds—the baseball player may also be an Exxon employee, a family man, and a devout Jew—and through the concepts and symbols in which their meanings get expressed. (Even the planetary-scale HTX is related to the yet-larger world of basic human common sense because it depends on human beings to exist.)

The Big Game: When Did the Global World of the HTX Begin?

The HTX enfolds millions, even billions, of our little worlds. But what about its limits in time? My students always want to know "when did the HTX start" or at least at what point it is unmistakably recognizable as the HTX?

This question can be misleading as it tends to exaggerate the importance of *epochal breaks,* like the start of the era of steam or the end of the Imperial Age with World War I. As some essential elements in the HTX have their roots back to the earliest prehistoric technical discoveries and to the first development of spoken language; as lasting essential contributions have been made from the first conquering of an energy transformation process (the ability to light a controlled fire), while the potential of new developments of essential importance are just now dawning on us (for example, the CD-ROM world of education and "interactive TV"), it is inadvisable to draw too sharp a line and say,

"Now after this point we are beyond primitive industrialization into the truly planetary system, with a new set of managerial notions and new capabilities for deliberately creating cyberspace which we recognize as pure HTX." One thing is sure: By the time there appear true planetary-scale corporations (before there was only the Church); a United Nations, with all its service organizations; faculties of business confusing managements with bogus concepts; and the "Internet" linking all "Thinkniks," the present "new" HTX world has for sure emerged, the Prigogine phenomenon of sudden transformation out of a previous chaos into a new order has happened.

This said, it is important to emphasize that within any new epoch, including the new HTX, there remain important elements spawned in earlier epoches. Much remains of an older industrial style, but it is under pressure to change, and mind-sets like positivism will linger long after their limits should have become recognized. Other elements, of ancient origin, retain a validity across eras of vast epochal change. Some, like writing, will cease to be valuable only at the end of man. Falling for "new is better" is very HTX and as silly as failing to recognize the great utility of what has just become available.

The only way to get the measure of this interweaving of processes, the acceleration of change, and the emergence of new dynamic HTX forms is to retrace their course of development, "to pull the bow as far back as possible," as Buckminster Fuller explained when asked how he saw so far into the future.[5]

Because the HTX, and so our situation in the world, is a historical phenomenon, we must learn how to think, not just process as such, nor the dynamic, and the interweaving of processes, but conscious *human development*—Postman is right about that: What has already been accomplished, as a storehouse of possibilities, is knowingly as well as implicitly drawn on by human consciousnesses to furnish a pro-ject of an aware, activistic, sought-for future. There is nothing more human than new being deliberately striven for but built on the possibilities handed us (tradere = tradition) from the past, which we then, ungratefully, irreligiously, take for granted, at the risk of eroding them, destroying our nurturing roots.

You do not have to be a fan of recent French sociopsychology to see

---

5. I was present when Ingrid Leman asked "Bucky" that question at a small gathering for his eighty-fifth birthday.

the centrality in human nature of *desire*.[6] Recall the wise word of St. Augustine: "Show me a man's loves and I'll show you the man." We strain so much to realize our future (especially in the HTX: it is so "Western" because rooted in the Judeo-Christian revelation that mankind is called to a destiny, a fulfillment), we easily become forgetful of our debt to the past.[7] Postman goes so far as to claim Technopoly destroys tradition. W. T. Anderson agrees: the tradition loses its roots and becomes not *the* authoritative voice but an authoritative voice, so that now we pick and choose (hence the complaint of the Pope against "cafeteria Catholics") what parts of the old truth to make (in a techno-logical sense in which knowing and making interpenetrate) into a ground deeper than the *Grund* of the tradition itself.[8] That has many serious practical consequences: Failure to cultivate the seeds from the fruits of old gifts leads to the disappearance of a valuable species. For instance, we allow the language to degrade. Or the sense of the rule of law fritters away or gets perverted into litigiousness (Hegel, in the *Philosophy of Right,* considered that the surest sign of societal deca-dence). Or the very sense of divine revelation evaporates.[9] Worse yet, the seeds can be transformed, reflecting facets never intended for them. (For instance, eschatology—concern for the last things—is dan-gerous when we forget its revelational origins. The Jews endlessly re-peat, "Remember what Jahweh did for our ancestors." Christians at the Eucharist repeat Jesus' command at the moment of consecration of the bread and wine: "Do this in remembrance of me!" The HTX—our collective consciousness—slips its moorings by not remembering, by lacking gratitude, and so no longer *listening* to the still-valid voices from the past). The loss of time implicit in the erosion can cause a crash, pri-marily because painfully learned lessons (or even revelations) are for-gotten. (The other night, I watched a television program on which critics discussing the $200 million flop *Waterworld* agreed that Holly-wood forgets hard lessons "within five years, because the big egos are always driving blindly forward.")

6. Superbly examined and commented on by Eugene Webb in *The Self Between: From Freud to the New Social Psychology of France* (Seattle: University of Washington Press, 1993).

7. To see the extent to which this is true, and that "history began at Sinai," the great work of Eric Voegelin *Order and History,* vol. 1, *Israel and Revelation* (Baton Rouge: University of Louisiana Press, 1956) shows it admirably.

8. Anderson, *Reality Isn't What It Used to Be.*

9. If just myth spun by primitive imagination, that may be a gain. But if there is a Word of God, which most of mankind believes there is, perhaps disastrous!

But it is not an erosion of the base per se, rather it is the transfer-ence of the shrinking of time and space in high tech to *a misconstrued notion of the nature of the cosmos,* as though it did not require a human deployment of time. Notice how provincialism rises alongside the free-ing of space implicit in HT travel and communications. To change the metaphor from the horizontality of the historical to the verticality of a hierarchy: Erosion at the base—the individual incarnation of human nature—can bring the whole towering "upside-down pyramid" struc-ture of HTX society crashing down, what our group calls "the earth-quake phenomenon." Human development can only be guided healthily if the principles directing it are well grounded both "horizon-tally," historically, and now in the "vertical axis," planted in nature and open to "transcendence from above." Those guiding principles must be not just "from the tradition" but soundly seated in all of reality.[10]

10. This is a central theme in the work of Eric Voegelin, explored throughout the five volumes of *Order and History.*

Chapter 2

The Seeds of the HTX

From Wandering Hunter to Cultivated Tribesman

Armed with the ability to warm himself and cook his food, and, eventually, with better weapons for the hunt, and then language allowing him to reflect on all this, why did man take such a long time—hundreds of thousands of years—to reach the stage of settled agricultural life, and then but a short few thousands of years to prepare the explosion of "civilization"? The invention of civilization was the greatest leap forward until the industrial revolution, almost comparable to the recent HTX explosion. Obviously, factors beyond the important ones of his tool, energy transformation and symbol technologies, were at work.

Adopting the "necessity is the mother of invention" hypothesis, we can assume that competition with predator animals spurred man on. At first weaker man would have scavenged the carcasses from their kills, but when driven to hunt for himself, he would have had to overcome his slowness and lack of physical strength. He had to perfect instruments and tactics. It is true that many inventions result from a bright person noticing a chance occurrence, but *when chance is too dominant a factor, much time is required for it to produce important results.* Chance still operates in the highly analytic HTX, but we do not base our planning on it. The very complexity of the HTX networks increases possibilities for chance suggestions to pop up, and good ideas travel fast over the Net. One of the strengths of the modern West has been its lack of uniformity, which allows easy access to new ideas and new faces. The relative chaos of America remains a strength as well as a weakness. The hunting tribesman lived in the midst of a drab, almost static uniformity.

In primitive times, these chance innovations must have sufficed for the sparse human population. And the nomadic life of hunting-gathering groups, who found shelter in caves and subsisted on a few grasses, berries, and meat, did not favor much of a population explosion.[1] So

---

1. Architecture arises only when man settles down to cultivate fields. Architecture belongs to culture, which raises an interesting question: Is today's in-

herds and tribes stayed in equilibrium for hundreds of thousands of years, language and storytelling advancing, step by poetic step. Without writing, storytellers developed prodigious mimetic powers. But the myths do not have much impact on the hard realities of daily life. Indeed, perhaps they distract man with visions of an imaginary world (the mother of "virtual reality"[2]), with roots in the soil in strange ways that skewer *cultura* away from reality while providing a vehicle for reflecting on it. Moreover, a herd of humans cannot carry much "cultural baggage" when it is always on the move.[3]

Supposedly it was depleting wild herds that slowly brought about the inventions of domestic herdsmanship and the selection of plants for slow improvement and cultivation. The transition from nomadic herd to settled tribe demanded new social arrangements. Where before the little band of migrating humans rather passively followed the cycles of nature, now a tribe accumulated a past in this one place, including cleared and improved fields, huts, and water supplies. Man began to have a home, and he had to plan ahead, at least for the year, in more detail than before.

The Basic Conceptual World of Primitive Man

Primitive man lived in a spirit-inhabited world yet remained sensitive to the moods of nature and of the animals with whom he shared a destiny. How strong was the effect of the "protovirtual reality" mythical dimension in the lives of primitive hunter-gatherer tribes? We need to reflect on this to understand our own complex relations with reality. Human knowledge is inherently creative of imaginative worlds: This starts with the most basic conceptualization, as it pulls us away from the immediate impinging of *these* concrete things in *this* setting to re-

---

ternational HTX "architecture" not something essentially different from what is meant by architecture in the context of cultures and civilizations?

2. The present flight into virtual reality also distracts masses of people from more earthy realities and hence increases still further the component of unproductive fantasy in our society. This diminishes proportionately the energy applied to more fundamental endeavors, including promoting efficient production, founding lasting families, attending to civic duties, and sound religion. I shall explain later what I mean by "sound religion" as opposed to an HTX-inspired flight into increasingly groundless imagination, "New Age" surpassing the myths of old in inviting us to float aimlessly above reality. When reality is discovered to be exciting, the incentive to escape is lessened.

3. Close to home, compare the much richer culture of the agriculturalist Cherokees, who even learned to write, to that of the nomadic Plains Indians.

late us to a wider, intangible, partially imaginary reality. This occurs through the medium of a generalizing abstraction: I see not just this tree present here but an exemplar of the MacIntosh apple type. The resulting concept, while often reflecting real forms found in external reality, exists only in the mind, the real form itself exists in the world only incarnate in, and limited by, a unique set of concrete circumstances, much of which is left behind when the concept is abstracted. To help us understand later the wilder inventions of "virtual reality," we shall pause a moment to think a bit about how the abstractive mind works at the base.[4]

Abstraction is a flight from immediate reality in order to manage a broader reality. Consider the difference between my observing the MacIntosh apple tree in my backyard presencing in all its splendor and my recognizing it as an "apple tree." In its immediate self-presenting to my senses, it appears in a precise setting offering a mass of in-itself structured data. Standing out against a background of other vegetation and, farther back, the neighboring house, it is at this season not yet a blossomy delight but a sprout of bare branches, ending in sky-reaching shoots from last year's growth, rather prickly in all. The concept, "apple tree," which upon the tree's presencing surges up spontaneously from my treasure of previously processed data (become "information"), is attachable to this presencing thing because the characteristic shape (*Gestalt*) it offers calls for this re-cognition. But the original data that has over time been processed into the general informative intelligibility of my concept, "tree," includes contact with many kinds of trees, in different seasons' arrays, almost all of which reality is itself no longer explicitly held in the notion. The abstract concept "floats" above them all. The concept belongs to an inner, spiritual world of consciousness, I can play with it imaginatively, turning it this way and that, more flexibly than the best CAD-CAM program. When I want to exit fantasy land, however, I can always refer my idea back to fresh data coming in from observation of trees.

To move from abstract concepts to the weaving of myths, primitive man built a more elaborate protovirtual reality inner world to deal with the larger questions of *the sense of it all:* we call this world mythical. Myths address large-scale practical questions too, not just theoretical-strategies, one might say: What do our lives mean? Where are we

---

4. What follows is necessarily summary. It took me several hundred pages in *Being and Truth* (Columbia: University of Missouri Press, 1996).

headed? How are we to placate the great forces? In myths, observations of larger forces and bigger issues were woven into stories in which the gods and nymphs and heroes became real for the storyteller. These stories enfold both something of the accidents of dynamic existence and the feel of order in a cyclical seasonal world. Mixed in with this effort to manage reality was magic, techniques for influencing the Powers. (Civilized modern man also has his myths. But there is a significant exception in all this: the myths of the Old Testament have a basically antimagic, antimythical thrust, for they stress the one God's absolute sovereignty and man's obedience. Believe it or not, this startling exception proves basic to the development of occidental technology mind-sets. We shall see about that later.)

Slow Primitive Evolution

Why did the primitive world evolve so slowly? Some parts of the world still enduring today are largely unchanged after a million years. And why was civilization so suddenly creative, when it at last exploded onto the scene some five thousand years ago? In the answer to this question there is a clue to the present still more explosive growth of the HTX.

For development, a certain kind of "imaginative reality" is required. Too absorbing an immediate contact with the directly accessible reality prohibits development. But certain kinds of "imaginative reality" (I shall call it "unproductive myth," not fertile in advancing technology and economic organization), and other forms of flight into imagination, lead the imaginative into "lotus land," dulling the evolutionary appetite of the masses: hard work in civilizations dominated by those kinds of myths then seems only for the dullards.

Consider, for instance, the hard e-ducative (*ex-ducere* = to lead out) work necessary to pull children up into the disciplined, hierarchical, and goal-oriented world of highly developed adults. It is the struggle of long-range planning against immediate impulse (and today against escape into television art forms, nihilistic music, sexual adventure, and that most dangerous art, drug and alcohol abuse, where one ceases even to care about reality). Then having succeeded in educating the child to a larger vision, what do we find: too often the overly developed adult HTX "voluntarist" whose life is so ruled by longer (read medium) range goals that he can never be settled in the moment (at a cocktail party he is always looking over your shoulder to see if there is not someone more usable in the crowd; in his car, he is always interrupting his conversation with a passenger to phone somebody). He is not receptive to the rich presentation of the real world surrounding him; he is so cut off from nature, he has nothing to contemplate, and his interest in human being is largely functional.

Primitive man was like the child, absorbed in the play of the moment, so immersed in the routine of daily tasks, and so attuned to nature, molding himself to its demands, that he allowed his longer-range tasks all to be set for him, broken down into traditionally regulated activities. Not much to disturb the millennial, seasonal, and daily rhythms he observed. He was thoroughly rooted, too immersed in low *cultura* for development to be quick. In the long darkening evenings, he naturally dreamt of another world. . . .

The Birth of the First High-Tech Worlds: The Advent of Civilization

But look what happened with the explosion onto the scene of civilization. It is important that we understand this, for the HTX, which grew out of Western European Christian civilization, is now an overlay on all the civilizations contemporary man has inherited. We need to understand how the HTX relates to its mother civilization, whether it remains at all dependent on it, and what it is doing in penetrating and transforming a vast variety of cultures and the several surviving civilizations, and whether it is in turn being affected by them.

Progress in agriculture in the fertile river valleys produced a growing population, putting pressure on the narrow strips of cultivatable land.[5] In the case of the Egyptian and Sumerian cultures, hemmed in by the desert, it is fairly clear what happened. Necessity forced the first large-scale ("systemic," we might call it) technological innovation: irrigation—the transportation and distribution of a natural resource over a large area, on such a scale as to demand a new level of organization, an administration, vastly exceeding the family's or tribe's management of its fields and herds. The development of a more dispersed feudal economy in China and along the Indus is less clear.

The radically altered social structure of the Egyptian and Sumerian civilizations can be glimpsed in its different criteria for success and its amazingly innovative symbol systems. As canals had to be dug and maintained, and water had to be meted out following some scheme, it became necessary to measure and to keep records, so specialized workers could be compensated with some of the fruits of the agricul-

---

5. Along the Nile, Tigris-Euphrates, Indus, and Yang-tse Kiang, civilizations developed at roughly the same time. This rough simultaneity remains a mystery.

turalists, the ultimate source of wealth. Success was no longer limited to the amount of game killed in the hunt or the abundance of the harvest. Now it included keeping the canal system working, distributing the water "fairly," by some socially acceptable criterion, and seeing to it that the specialized workers get paid.

For all this to be possible, an amazing new kind of symbol system, money, had to be invented, much more flexible (and abstract!) than barter.[6] And a better way of remembering "data" ("data" now becomes *explicitly* important) was needed to keep track of basic production, the channeling of natural resources—water, in this case. Also required was higher technology energy transformation (the forced draught forge) allowing smelting of metals, fabrication of metal weapons and tools, which, in turn, permitted stone architecture and heavily armed troops— a military with capital investment in weapons and social investment in training to protect the city and the canals. The bureaucrats of the kings of Upper and Lower Egypt and of Mesopotamia would have found computers handy, but they started with the prior symbolization, spoken language, added a new kind of analysis, mathematics (geometry and arithmetic) with its own new symbols, and then invented ways to record all this by engraving it on things—mud bricks in Sumer and papyrus in Egypt—that could endure "out there" in the cosmic time-space of material reality. These inventions, writing and mathematics, were man's greatest technological feat after spoken language. (But they transcend technological use by virtue of their poetic character—what is more beautiful than a superb mathematical proof?)

Control of the medium of writing gave power to the bureaucracy, the clerics (clerks), who also fostered among the poets development of a new civilized "cosmological" mythology, adroitly forged to ensure the imperial power.[7] (When the HTX begins to become visible to the modern poets, the mythology moves from the civic realm to become an ontological claim: the myths of men like Hölderlin and Nietzsche. And the busybody imagemakers are still with us, amusing the masses and suggesting to them what they should believe, a vital part of the power

6. Caution! The development of money was not simple. *Encyclopedia Britannica* claims, "Money is to be thought of as originating out of religion and social custom rather than directly out of barter. Commodities which later came to be important as a standard of value or medium of exchange seem first to have been important as ransom, ceremonial offering or means of ornamentation" ("Money," 1966, vol. 15, p. 702). This only makes clearer just how important the symbolism of money (borrowed from cult and ornamentation to serve economics) really is.

7. Voegelin documents this in a profound way in *Order and History*, vol. 1, *Israel and Revelation* (Baton Rouge: University of Louisiana Press, 1956), part 1.

elite's manipulation in HTX-transformed civilizations. From the first more or less conscious manipulation of the masses through myth-making in the early civilizations, through the mythologies of the "Sun King" [Louis XIV], the *liberté, égalité, fraternité* ideology of the "Enlightenment," and the even more controlling ideologies of the to-talitarian states, to the politically correct and greedy Disney mythmak-ers of today [*libérté, frivolité, latté,* said one wag],[8] part of the being of every epoch has been rooted in a voluntaristic creation of the con-cepts and symbols of civilization).

Because human beings are bodies, *geographical space* remains a fun-damental factor in civilization. Globally, its importance is being weak-ened by HTX transportation and communications, but locally, ground is becoming more crowded and expensive. Indeed, civilization has been defined by geographically induced challenge and response.[9] Now in the HTX, humans are dreaming of going beyond all natural con-straint, with advance to other planets and genetic engineering of bod-ies for longevity and durability, able to make the trip into the cosmos. But dream it remains in a world where geographic and life-time con-straints have not disappeared, and never will.

Imagine the being of the great early city. Here "being" is not merely physical reality but the whole imaginative weave of notions about the city, the mystique about it and the know-how to live in it. The new cities so impressed the poets, the biblical Jonah story declares that Nineveh could not be traversed in three days! Lived space being an il-lumination of being, that is, the projection of a group living together imagining and sharing it, the way a tribal person would relate to the new spaces and structures of a Nineveh is very different from the way a modern New York archaeologist would view its ruins, extending less widely than Central Park. In attempting in a contemplative mode "to dwell in the being" of the great city, all of the following have to be imagined, and related to one another in one "world of (civilizational) discourse":

8. Maximillian Robespierre Meese, a cat on the island of Unalaska.
9. Consider, for instance, the impetus to colonization on the part of a defor-ested Britain, requiring trees for both ships and fuel. Not only did this force accep-tance of coal, with all that entailed in the impetus to industrialization, but it led Great Britain into disastrous handling of the American colonies, not least the proclamation of the Québec Act of 1774, which cut off the trans-Appalachian areas from settlement by the thirteen colonies, preserving its resources for the Crown, with the colonies then retaining an increasing portion of their timber for their own consumption.

—*large scale projects,* like irrigation, temple building (architecture—a whole new set of symbolic forms—first appears), city planning (there is much that is symbolic in the laying out of the city, from Memphis, through Major L'Enfant's design of Washington, to Haussman's Étoile, with the grand boulevards radiating from the Arc de Triomphe, to Brasilia, but now most cities just sprout in a utilitarian fashion), all of which requires a new kind of symbolization, including the elaborate mathematical calculating-planning of engineering, which itself only becomes an explicit concept in the eighteenth century, four millennia after it became a reality;[10]

—*specialization*—engineers, canal diggers, water-regulators, architects, metal smelters, metalworkers, merchants, caravan drivers, moneylenders, priests, scribes, higher administrators, all but the agriculturalists and herdsmen now lived crowded together, in order to be readily accessible; the social fabric of the society becomes many-hued and more hierarchical; hence

—*hierarchy and bureaucracy* (with the concept of "office," and new symbols of authority and prestige, including elaborate dress, and a multiplication of cults and political games);

—*transfer of fruits of human efforts* among specialists—capital surplus from agriculture (the primary productive domain) diverted by fees for water, taxes, temple tribute, to other specialties and to governing and priestly hierarchy. While much of the payment may have been still "in kind," this is cumbersome; that is why

—*money* had eventually to be invented. That took centuries, and it did not uniformly endure—the Eastern Roman Empire eventually abandoned money as cash and coin and returned to in-kind payments. These shifts of financial concept are important, because they are all-pervasive and assume great symbolic power. As we shall see later, the HTX has developed a new sense of the symbol "money." (Better conceptions of "capital investment" take until the eighteenth Christian century to develop). This provides a good example of how further conceptualization marks an important illumination by Being.

—*writing* and record keeping (imagine the dialectical development—growing complexity of the society requires keeping track, invention of primitive systems for doing this increases the ability to keep track of

10. What is it about the HTX that brings explicit symbolization forward so rapidly, belonging more to a science fiction novel than to daily reality? Often an imaginary world will be built and names found for things which do not exist yet, and may never come to be.

larger systems, which in turn require better accounting, but all presupposing a basic sophistication in the overall illumining by Being);[11]

—a *mythology* more in keeping with the hierarchy, scale, complexity and specialization of the new society, and its governing structures, the more elaborated "pantheon of the cosmological empire," as these more sophisticated priests struggled to make sense of an ever more complex world, and to aid the authorities to control the masses;[12]

—a centering of the new order around the city, dominated by the most impressive of all the symbols, its great *permanent stone temples and burial monuments—like the pyramids and the magnificent palaces,* all leaping many orders of magnitude in size and complexity beyond anything previously built by man. The village is forever marginalized, although, three millennia later a majority of mankind still lives in villages. But then the banks were until recently still building eighty-story temples to the god Moolah.[13] (Are skyscrapers still appropriate symbols in the new HTX, or is the casual campus of a Microsoft in Redmond, Washington, not a better expression of the present reality?)

All this interweaving innovation has to be managed in the imagination, hence the essential role of symbol innovation and manipulation throughout: the mathematical calculations of engineering, the concepts and prestige symbols of office, writing and record keeping, money, governing and meaning-giving mythology, and the elaborate palaces and temples with their carved symbolism. *The more complex society becomes, the more man lives in this interior symbolic space, and steps toward ever more explicitly elaborated "virtual reality," the stronger grows temptation to prefer imaginary to lived intersubjective.* Nature-rooted space is transcended into imaginary-creative being, a different kind of escapist mythology. (In the new HTX, entertainment—including tourism, a carefully managed way of floating through someone else's space and culture—is the largest industry.)

11. It is often asked, which came first, division of labor, leading to political organization, or political organization empowered by securing and distributing a surplus, leading to division of labor? It will prepare us to understand the weblike essence of the HTX if we suppose that it was—like writing-managing—more a both/and, a continual dialectical ratcheting up of complexity due to a constant interplay of both productivity and political organizational factors.

12. This latter is not, of course, the raison d'être of priestly activity but a perennial temptation and perversion of divine office.

13. The Greeks coined the term *archi-tect:* he is the "fundamental maker." The new civilizational Sein is well reflected in the term. Interestingly, some civilizations did not build in stone until they were growing old—the Greeks and the Hindus are cases in point. At their height the Greeks were imitating their earlier wooden temples in stone!

With civilization, *social classes* emerge, providing a much greater separation than that of elders from the rest of the tribe. The classes become quickly so distinct one can wonder, do they all live in the same "world"? In some respects, yes, they walk the same paved streets, they see the same great temples, the palace, the open shops, the neat fields; they all know this is Thebes, and Pharaoh of the Double Crown is King of Upper and Lower Egypt. Because they all possess objectively the same human nature, they can all acknowledge the same objective giveness (the esse, the existence in cosmic time-space, with a particular form, regardless of what anyone may think about things) of the great city.

But the lower classes are not admitted into certain spaces reserved to the elite, and the different classes begin to speak different languages, at least distinctive dialects. Whereas the priests and clerks can progress in handling concepts and measures by writing down symbols, much that the learned understand and talk about is "in a different world" from that enveloping the concerns of the petty merchants and the slaves. Records of what went on in the past gradually come to exist, which the learned can read. Because the lower classes are illiterate, their past is restricted to what they can recall. But, as we shall see in a moment, neither class yet has a full-fledged *eschatalogical* sense of history, that conviction that mankind (as a whole) is coming from somewhere and pursuing a destiny, that humans (equal before God) will arrive at a fulfillment.[14]

The horizons of the literate are much greater. The higher classes in the early civilizations are less exclusively absorbed in the routine tasks of every day. They are party to the elaborate routines of court, still built around the seasons of the year (despite the political and cultural dominance of the great city, these remain economically agricultural societies), and they are the elaboraters of the great liturgies of the "cosmopolitan empire," celebrating those cycles of the seasons. They are the principal authors of "high culture." (Cultures become "higher" in direct proportion to the transcendence achieved through increas-

---

14. "*Istoria*" (researches) is what Herodotus, the first Greek "historian" called his inquiries into surrounding civilizations, which he related and contrasted with his own Greek civilization. More than a half millennium passed from the beginning of written records, followed later by laundry lists, inscribed on great stone monuments, of events of the great kings, until Herodotus's full-fledged researches. But even the Greek historians lacked the Hebrew's sense, which began at the event on Sinai, of a story of humanity marching from a meaningful, willed beginning to an ordained fulfillment at the end of time.

ingly explicit critical reflection. Think for instance of the progression in insight and abstraction from gestures and grunts, to spoken words, then grammatical discourse, invention of the alphabet, writing, and finally the use in our time of Boolean algebra to make machine reproduction of every possible letter and symbol and image possible.) The lower classes follow these goings-on of the higher classes passively and from afar, the priests actively and from within. (Later, we shall ask who are the "high priests of the HTX," and how do they see the world differently from those lower down whom they manipulate. The growing gap between the high-techies and the drones gradually being replaced by computerized machines is a serious new social class problem in the HTX).

Even for the highest classes the horizons of the cosmological city-state were, from our perspective, narrow, at least horizontal. Vertically, sights were raised to heaven (via the Son of God,[15] the supreme intermediary sitting on his magnificent throne, center of the great festival sacrifices). In the great City, there was no question where the *umbellus mundi* was to be found. (In the HTX the horizontal spread to all the planet, with its inevitable busyness, distracts from the vertical looking up to heaven, and there is no umbellus! Just as each of the cosmological empires considered its own capital the navel, and so there were several, so today, despite our images and voices being present almost anywhere electrically at the speed of light, the political-social nature of man still sees him clustering, "eyeball-to-eyeball," negotiating, socializing and being in the right circles. So while none of the "world capitals" would seriously claim to be the navel of the world, several do, despite the uncentered nature of "the web," exercise a certain dominance.)

It would be worthwhile to follow in detail every advance from civilization, through the great "ecumenic empires," to the emergence of national states in the midst of the medieval Holy Roman Empire, observing technologies, social structures, and symbol systems interacting, as the epochal being evolves, the technology complexifies, and the political economy develops. It would also be profitable to contrast the development within Western civilization with the quite different preindustrialization course taken by the other great civilizations. Both such studies would prepare us better to appreciate what happened in the in-

---

15. Many are unaware that this term was used for pharaohs and even of King David. Jesus was cautious in applying it to himself, preferring the prophetic "Son of Man," although when Peter declares his faith, "You are the Son of the living God," he does not reprove him.

dustrial revolution and the subsequent development of the HTX. But we have so much to do just to sketch the recent exploding onto the scene of the HTX, I have to jump straight to its threshold, the "industrial revolution" of the past two centuries, and even then I shall offer only the most sketchy outline of that rich story. The few hints I have already shared about the early civilizations will suffice, I hope, at least to continue the discussion of the interaction of technological innovations, revolutions in information processing, "ekistic" structures with theological-philosophical (mythological) conceptions of the world, the HTX forms of which are the central concern of the present volume.[16]

From Civilization to Industrialization

From Agriculture to Industry

In considering the emergence of the HTX, we must first ask: What changes of mentality reinforced one another to produce the particular (and unprecedentedly dynamic) reality of *industrialization,* that mother of the HTX?

I propose, in addressing this question, to pay particular attention to the technical breakthroughs and shifts in scientific worldview that are involved. (The Cartesian-Newtonian worldview is intimate with industrial organization, just as is the Einsteinian-complexity worldview with the HTX). Granted, we live in a postindustrial information society, nevertheless it obviously retains an important industrial core, just as we still depend on an agricultural dimension (albeit also industrialized) as provider of essential foods, fibers and organic molecules.

The Dynamic Dialectic between Being and Thing

As we look at the interweaving of certain processes in the first industrial society, Great Britain, and at the invention of new cultures and institutions to control them, now called "industrialization," we shall see why each process is necessary and how they all feed on one another. Nothing shows better how the whole interacting complex is illumined by the developing light of being—the evolving overall sense of what is going on—and reciprocally that collective being-interpretation of the emerging industrial world is receiving "incarnational substance" by those processes, transforming actors and things. Behold "*the dynamic being—thing dialectic*"!

---

16. "Ekistic" is the Greek architect and planner Constantinos Doxiades' short term for the sociopolitical-economic reality of all human settlement.

Out of Persisting Ancient Roots, the Radically New

From the classical source, the seventeenth-century savant, Gregory King's estimate of the population and wealth of England and Wales for 1688, we learn that the total population of England and Wales on the eve of industrialization was 4.5 million.[17] Contrast that with the 60 million or so the HTX allows to flourish today in the same territory: 4.5 million multiplied by a 38–year life expectancy equals 171,000,000 man years, versus 65 million with a 75–year life expectancy equals 4,875,000,000 man years, supported by HTX ekistics, with lower criminality (index of good order) and better nutrition to boot! That is roughly a 35–fold increase! Of 1688's 4.5 million, 2.8 million constituted the "laboring poor," folk needing assistance from private charity and institutional poor relief. This arose from the "enclosures," the "industrialization of agriculture (phase 1)."[18] In the words of Peter Mathias, whose account of the English development I am following here, "The frightening scale of the problem of poverty revealed by Gregory King is paralleled only by non-industrial, underdeveloped countries today."[19] Underemployment, and low productivity, were the universal curses then that they are now in "the Fourth world."

Who in the then existing social groups were able to save money, and how much? These were critical questions for the *formation of capital* needed for the eventual industrialization. King estimates the annual savings of freeholders as £560,000; merchants, £400,000; landed classes, £400,000; farmers, nearly £200,000; persons in law, £210,000; and the depressed half of the population, nothing. Thus the main source of savings was the rural society. As people acquired wealth, they tended to invest it in land and then live off their rents. Most other European countries had more serfs, with a greater degree of subsistence. England and Wales were already highly "capitalist." Foreign trade already played an unusually high role, only Holland was comparable: about 10 percent of a net national product of 50 million pounds was imported, with exports at about the same level.[20]

All of this preindustrial manufacturing, commercial, and agricultural activity took place, of course, within a certain civic order—law and

17. Cited in, and analyzed by, Peter Mathias, *The First Industrial Nation: An Economic History of Britain, 1700–1914* (New York: Scribner's, 1969).

18. Seven million acres were covered by parliamentary enclosure awards between 1760 and 1815, with over one thousand acts passed between 1760 and 1800 (Mathias, *Industrial Nation,* 73).

19. Mathias, *Industrial Nation,* 27.

20. Ibid.

government were well established. Mathias stresses *the importance of a stable political regime and a rule of law,* including patent protection, for industrialization.

Taxation and public expenditures in relation to national income were lower than in other European countries which accepted greater public commitments. Natural, spontaneous economic flows were less distorted; less private capital was siphoned off. In particular, lags in social overhead investment by the state in the first half of the nineteenth century affected the process of growth. While not making demands upon investment resources, they increased the social costs of the transition to an industrial, urbanized society. Compared with Holland, the impact of taxation on the level of demand (or on interest rates) was significantly less.[21]

In the landed economy, Mathias declares, "social flexibility was reinforced by mobility in the social hierarchy over one or two generations."[22] That is misleading. The underlying point is true: *Inaccessibility to capital and the tightly closed "aristocracy" of countries like Colombia to this day hold back the pace of industrialization in resource-rich lands. The "open-endedness" of the aristocracy, as just described, was important:* Because of primogeniture, all but the first sons were pushed out into the professions, or even into trade. Wealth easily bought "knighthoods, rotten boroughs, and a parliamentary career."[23] "Commercial instincts learned in trade brought to the land habits of accounting and profit calculation learned in trade, *habits of ploughing back capital into a business to expand it,*" a good illustration of the illumining of the epochal economic scene by being, in the sense of a widely shared set of notions.[24]

Farming being a business in England more than a subsistence pursuit to support a peasantry meant that output per head in agriculture was maintained and increased at a time of increasing population. Underemployed marginal labor on the land was being squeezed out as numbers rose. Social pressure on the land thus brought economic advantage to the country, particularly in rising industries that had higher output per man than agriculture. The expansion of national wealth created in this way brought, eventually, a long-run social advantage to everyone, compared at least with the social devastation that came with famine to the peasant societies in Ireland and the Highlands in the 1840s.[25]

21. Ibid., 43
22. Ibid., 52.
23. Mathias, *Industrial Nation,* 53.
24. Ibid., 53–54.
25. Ibid., 64

Shifting and investing capital demands a strong banking system. Because industrialization is fueled by capital, a strong banking system was essential: landowners and successful farmers banked their surplus instead of hoarding it at home. A national system of banks was then able to gather significant sums and direct them into transport infrastructure and into industries.

Development of such a banking system required the gradual creation (starting in about the fourteenth century) of a whole new set of concepts, manipulated through a growing symbol system, "letters of credit," "term deposits," "coupon bonds," "spread,"—the list is endless, and it continues today, growing so fast the average person is bewildered by all the "savings instruments" spread out gaudily before him in the bank's advertising!

It is not only capital that has to be shifted about, population must too, but then people have to be fed. An increasing industrial labor force coming at a time of growing population creates the need for increased agricultural productivity or for importing of food.

Increased investment in land improvement, enclosure, and improved stocks helped in England.[26] Of these, enclosure was the most important (and the most brutal), because it made all the other innovations possible.[27] One can generalize even further: *private property plays a great role in industrialization,* because—think of human selfishness what you will—this sense of "it is mine, I can do with it what I want, and I will enjoy the fruits of my labors" is a powerful incentive for work and risk. Perhaps half of England had been enclosed by 1750.

Transport: Fueled by Capital

"The high capital costs of investment in transport make transport improvements one of the potential blockages to the progress of industrialization. . . . A large transport project needs to be complete before its benefits accrue from tolls to pay for it. One cannot begin the venture in a small way."[28]

England enjoyed many advantages in this respect, as did Holland. The growing and profitable agricultural and textile sectors, and the existence of solid banks, provided the capital. But nature helped too: no

26. Ibid., 66
27. Ibid., 72.
28. Mathias, *Industrial Nation,* 115.

point in England is more than seventy miles from the sea, and some of the best coal fields were either close by the sea, or not far from navigable river systems. With the exception of a couple of strategic highways, roads were miserable, and the government was disinclined to invest heavily in this or any other form of transport infrastructure. But a largely unplanned network of local turnpikes began to be stitched together, and in the last half of the eighteenth century, as coal both for heating and steam power in mills began to be in demand, privately built canals connected north to south the river systems that flowed out to east and west. "Financing canals and turnpikes showed how plentiful capital was in eighteenth century England, and how favorable the social context was for investment."[29]

The Great Breakthrough: Steam Power

At the dawn of human existence with the harnessing of fire, a series of breakthroughs in man's *ability to steer energy transformation* began. Wind and waterpower were later also domesticated. But the biggest leap came a million years after domesticated fire—the middle of the eighteenth century—steam power, the innovation Andrew Grove, CEO of Intel, considers a greater revolutionary leap than microelectronics! (He may be wrong. Learning to control flows of elementary particles, the electrons of electricity, is arguably even greater than raising the pressure of and channeling steam.)

Steam power was the greatest of the technical innovations developed in the course of the industrial revolution because it became the agent and instrument for applying basic innovations in so many industries and transport. Everyone knew that the greatest strides in technical progress lay in finding ways to use the steam engine and iron machinery in more processes in more industries. A contagion was in progress. The supremacy Britain had achieved in technology by 1800 was also based upon a combination of cheap iron and coal, skills of accurate metalworking and the skills of mechanical engineering—in machine building and steam power in particular. Each of these was growing more dependent on the others. But the gap between Britain and the most advanced other economies in this field was quite unprecedented and unrepeatable.[30]

29. Ibid., 117.
30. Ibid., 134.

The sense of the all-englobing illumining "being" is palpable in Mathias's description. "Skills" in iron bashing, in steam harnessing, in machine design, and in mechanical engineering seem more telling than the well-placed supplies of iron and coal. After all, Lorraine and the Ruhr, to name but two, had those, and yet they lagged England by a half century. No explanation of the peculiar development of these mind-sets and attendant accumulating abilities—the particular "social capital" of Britain—is offered. But there are hints.

Enough managers had to *conceive the possibilities* latent in the developing techniques, and *conceive of ways of mobilizing the investment and the manufacturing techniques* necessary to make use of these possibilities—breakthroughs of understanding—leaps of *Sein*—commanding a whole organization of not just manufacture but marketing.

Expensive machinery and large plants increase fixed costs. Industrialists with expensive machinery therefore had rising incentives to keep their plant running as near full capacity as they could. The increased output increased the flood of goods, but so also did their new commercial incentive. Both led to the need to extend the market and to cut prices to sell more. These are commercial incentives which help to make innovation a cumulative process. *One innovation breaks an equilibrium in a traditional sequence of processes.* The flying shuttle created such a demand for yarn by increasing the productivity of weavers that it created a great incentive to increase productivity there. Hargreave's spinning jenny was born in a flurry of activity to do just this. On a larger scale, factory-spinning inspired power-weaving. Likewise, when a steam engine was attached to wooden machinery it shook it to pieces and required the innovation of iron machinery thus creating an incentive for a more powerful engine. This lay directly behind Watt's invention of the double-acting low pressure engine.[31] The modern chemical industry was born in Britain to remove a bottleneck caused by cottage industry sun bleaching. "*The disequilibrium could be one of speed, or of scale of production, of costs, of efficiency, of materials.*"[32]

"*One innovation breaks an equilibrium in a traditional sequence of processes.*" That sentence holds a key to understanding human development. The new matrix of iron-coal-steam-metal machining-railways gave increasing freedom from the old traditional limitations of nature—freedom from limits of animal and human muscle power; from dependence on water and wind power; the seasonal limit of the har-

31. Ibid., 141–42.
32. Ibid., 142. Emphases mine.

vest; precision machines freed from the inaccuracies of human hand and eye, creating the most "unnatural" standards of measurement and tolerances.

While vastly increasing the capital tied up in production, these innovations greatly *reduced the capital required per unit of production,* and they *reduced the cost of skilled labor per unit of production* as well. That is another key to the dynamics of modern development, true today and as long as manufacturing remains, which will be until the Second Coming.

But lower prices and wider distribution actually created new need for capital and for employment, and it bred the need for new skills faster than it destroyed old ones. *"The pace of diffusion of the new techniques was limited by the scarcity of skills as much as by scarcity of capital or problems of transport."*[33]

("Skills" obviously exist in the spiritual realm of human culture: mindsets and habits). This same restraint is operative in the HTX today.

Capital Accumulation and Efficient Capital Flows

Rich agriculture and lucrative foreign trade, plus a good savings rate, provided a steady flow of capital in the eighteenth century, as economic development unfolded rather smoothly.[34] The main challenge was overcoming social, political, and institutional constraints that prevented these savings flowing to where enterprise needed them.[35] "In part the problem was geographical, in part sociological; between the farming savings of East Anglia and the credit demands of Lancashire; between the entrepreneur in a fairly humble station in life and a landed magnate" (still a problem in the Colombias of today. Indeed "regional disparities" continue to plague HTX countries, from Italy's problem of the South to Canada's problem of the Maritimes. Nature and old cultures conspire to restrain development.)

Except for canals, docks, and deep coal mines, most industries needed much more short-term capital, to finance raw materials, goods in the pipeline, and receivables, than long-term, for buildings and machinery. This "working capital" was readily provided by local bankers and merchant credit, through the industrialist buying materials "on

33. Ibid., 144. Emphases mine.
34. The savings rate was fairly high despite much "conspicuous consumption" on such things as lavish construction, luxury clothing, and "Grand Tours" of Europe.
35. Mathias, *Industrial Nation,* 145.

tick."[36] Such an institutionalized development as the banker's intermediary role meant that "the inadequacies of depending on personal, face-to-face contacts to equate saving and borrowing were being overcome, so that *savings ceased to be largely hoarded, sterilized from productive use.*"[37] (This last marks a basic change of attitude, pointing clearly to a new epochal being, showing both an *abstracting and depersonalizing,* allowing "action at a distance"). The greatest source of long-term credit was the "plow back" of profits, as it still is today.

The entrepreneurial spirit continues to be a central driving force in the HTX. Who were the imaginative, ingenious, ambitious individuals with that spirit in the industrial revolution? According to Mathias,

> entrepreneurs sprang from economic opportunity as much as they created it. They depended everywhere upon a necessary creative environment. They join the circle of other factors in economic growth as part cause and part effect. . . . Latent resources can be unused until "men of wit and resource" organize them for a market they have promoted. *Once formed, attitudes, often consolidated within social groups, exert an inertia of their own, like institutions, for good or ill.* The industrial revolution was not merely consequential upon the economic logic of geography or geology.[38]

One of the entrepreneurs' prime organizing skills was *merchandising.* (Everything in the HTX has to be "hawked," starting with one's own skills. HTX = "Hawkers' Temperament Xtravaganza," or "Hype on TV to Xcess": it ain't just what it *is* that counts but how it is made to *seem.* Could a nineteenth-century medicine man have imagined the Disney hype around *The Lion King,* with a cottage industry of "Lion King" junk in unimaginably varied forms, from books to tote bags being sold with other "hype tripe" in specially erected stores? Mussolini [the super-innovator] and his imitators, Hitler and Stalin, hyped even more sinister ideological pap than that. Mass merchandizing of symbols out of greed [replacing the Church's slow, profound inculcation of symbols out of love for Christ] is a central HTX reality. "Lion Kings" fill a void, as spiritual junk food.)

36. In Russia today, there is too little new investment in modernizing machinery, due to lack of long-term credits. Working capital is in short supply, and interindustry loans have reached $15 billion, most of which will never be paid back!
    37. Mathias, *Industrial Nation,* 149.
    38. Ibid., 151.

The entrepreneurs needed to hold the initiative in purchasing raw materials, maintaining a *uniformity of quality* amidst an incoming flood of materials. Commissioned brokers came into being.[39] To achieve quality, the entrepreneurs had to discipline a previously untrained labor force, unaccustomed to bending to the demands of the engine-driven machines. (Education—more important, training organized and financed by the company—is one of the on-going engines of the HTX, the handing on (*tradere* = tradition) of being).

Consider the *character* that built industrial England. The economist Mathias is no romantic dreamer. Yet he had this to say about the role of the *faith* of these hardy men, mostly members of nonconformist sects:

> They maintained a strict insistence on personal probity and honesty. Luxury and idleness were the twin evils of these faiths, vices that presided, not only over damnation in the next world, but so often over bankruptcy in this one. Many Quaker diaries, like private confessionals, poured out details of the prices of the Funds and business deals interspersed with urgent warnings to their writers to remember the parables of the talents and the wise and foolish virgins. This is not just a question of thrift, of the peasant virtue of hoarding, but of enterprise and energy, of personal thrift and abstention in order to plough more capital back into the business and allow it to expand more rapidly.[40]

These entrepreneurs were big risk takers.

> The eighteenth century was a time of high risk and uncertainty in business, compared with later periods. The institutional structure within which business operated was weak and uncertain compared with our own day. Communications were slow. It was difficult to keep close control of agents at a distance. Legal processes for recovering debt were expensive, protracted, and less certain than today. The institutional ways of raising capital were very limited; so were institutional techniques for owning and managing enterprise . . . [after the] Bubble Act, each partner was fully responsible in his private estate for the depths of the partnership to the last guinea, to the last acre . . .

39. This is increasingly vital in HTX—75 percent of the cost of a microchip is attributable to testing—once you have made it *seem,* and thus sold it, it still has to work: a harder reality is always there at the end of the day. It took the Detroit of the 1970s a while to learn this ontological lesson. Another consideration: A company cannot afford to have high-speed, integrated, multimillion-dollar production lines grind to a halt because of a few defective parts.

40. Ibid., 159.

[Corporation status] was very rarely granted to a manufacturing industry . . . until after repeal of the Bubble Act in 1825.[41]

Result: "Kinship was the organizing principle of most business."[42]

An important accomplishment of the HTX is that it has made possible doing business with strangers and has extended vastly the range of trust.[43] The erosion of trust by organized criminal activity as in Russia today undermines an important good HTX effect. Indeed Russia's society lags in basic HTX spiritual development, despite pockets of state of the art technology. So entrepreneurs there have to be "big risk takers" exactly in the mode of industrializing England, but even more so, for they are without the Quaker probity to help. It will be interesting to see whether the sound conditions needed for healthy development will emerge from such a materialist, indeed nihilistic being as that of the core Soviet lands, a land without the Confucian kinship bonds that have so helped HTX development in Asia.

Restraint on Competition

As competition became murderous in various sectors of manufacture and distribution, a number of devices for "managing markets" grew up. Trade associations attempted to arrange prices and share regional markets. Cartels, formed through exchanges of shares, largely kept rather quiet in England while favored by government and banks in Germany, were often more successful. And finally—and perhaps most powerful of all—were the large combines, the *Chaebols* of the day. Either an "alligator company" would buy up its smaller competitors, or stronger companies would combine through exchange of shares into monsters. The success of J. & P. Coats in the thread business, Imperial Tobacco, and the Lever Brothers in soaps and margarines is well known. In England none of these so throttled competition as to merit the setting up of any government agency to fight them, until the Monopolies Commission was established by the Labour government in 1948! In Germany, where the economy was considered an ally of a strong state, the cartels and groups were often successfully used as means of achieving efficiency in production and in keeping the economy on an even keel.

41. Ibid., 162.
42. Ibid.
43. See Francis Fukuyama's study of this: *Trust: The Social Virtues and the Creation of Prosperity* (New York: Free Press, 1995).

Complexities of size of firm versus "efficiency," and of monopolies in various forms versus genuinely free markets (more a matter of degree) will remain as long as the HTX.

The Need in the HTX for an Industrial Core

This brief glimpse at a few aspects of the first century of the industrial revolution in England has amply illustrated how *attitudes* lie at the base; how subworlds, mutually interdependent, had to develop slowly to critical mass before "old equilibria" were broken and related developments occurred in other worlds; and how the "luck" of technical breakthroughs was also accompanied by dogged determination to reply to newly arising challenges through painstaking search for innovations. The strong role of a certain moral character has been noted.

All this remains relevant in the HTX, which, for all its "postindustrial" aspects, retains a base in human nature and in "industrial man." I would venture to suppose that, while not quite as perilous to the base as allowing moral character to corrupt, allowing the industrial foundation, especially manufacturing prowess, to corrode could be disastrous. That industrial core is the soundest producer of fiscal surplus, which remains the lifeblood of the HTX. Whether a given country's ability to produce foodstuffs efficiently is as important is debatable, but modern agriculture can be profitable too, if you look past the distorting subsidies. I shall keep stressing the solidity of the base, nature—human nature and natural resources—and of "the first and second stories," agriculture and manufacturing prowess, because, as we are warped off more and more into an increasingly fascinating world of imagination, capped by the wonders of "virtual reality simulation" the temptation of which I have already warned grows: *to build a colossal inverted pyramid on a base of neglected natural, personal and societal fundamentals.*

We knew first from revelation and now we have evidence from physics that THE world will come to an end. Long before the cosmic slide into entropy, life on this planet will be destroyed, as the fragile equilibrium that sustains it becomes too upset for recovery. We suspect that this will happen as a result of human intervention long before the sun's expansion to a red giant produces unsurvivable temperatures. As we contemplate the beginnings, the course of development with its leap to exponential growth in complexity and mass, and the predicted end, it behooves us to pay attention to ways in which we are weakening the upside-down pyramid at its base. There is not much sense in our hastening the end out of mindlessness and greed. We shall come back to that doomsday consideration in the last chapter.

In Mathias's analysis of the early industrial revolution, we caught glimpses of the evolving world of symbols and imagination: Increasing literacy, the invention of newspapers, the strengthening of scientific societies as instruments of communication, all played roles in that revolution. Why did a new art form, the novel, enjoy such popularity at the same time? Escape into a fictitious protovirtual entertainment world was not new. But perhaps since *Don Quixote*, a movement had begun to reorient this world away from the romance of chivalry toward the "real world" of everyday experience of a broader mass of people. Part astute psychological probing, rich in appreciation of psychosocial reality, the English novel of the eighteenth century is a modern, almost HTX phenomenon. The strong practical sense of those industrial and agricultural innovators was marked almost exclusively by a realism—a set of attitudes that kept their attention centered on the details of factual situations, on tough, real, immediate practical problems of mechanical innovation, market management, securing of resources, and quality testing. This invited little flight even into entertainment, let alone the forging of new ways to represent things. It was a very quiet world compared with ours!

I shall leave for another day speculations on the formation of the vast romantic symphony, played by orchestras with grand accumulations of man years of study (those years placed one after the other would, in the case of Mahler's Eighth Symphony, stretch back to paleolithic times: 1,000 performers multiplied by 15 years of study equals 15,000 years—an HTX extravagance! Only HTX-generated wealth and HTX management could produce the machine necessary for such productions.) And what an imaginary world to escape into! Musically invoked religion, HTX emotional therapy, and legitimate (I suppose) creation of beauty, all mixed together.

3

The Essence and Anatomy of the HTX

Introduction: On Social Form, Social Control, and the Use of Symbol

Is the HTX a Social Form?

Part of the core of the HTX, I have just suggested, remains industrial. But if we are to live wisely in the HTX, we must probe deeper to discover the form of this planetary-scale social reality, "the *new* HTX, world of the Internet," which has grown far beyond its Western civilization homeland and light years beyond the dimensions of mere industrialization.

The enormous scale of the HTX is new and demands new forms of thinking and vast organization. We shall examine these institutions to show how far the HTX has moved beyond the being of the industrial epoch.

Before trying to capture the form of the HTX, I should explain what is meant here by "social form." To do so, I shall show something about all social forms, ranging from those made up of family, local clubs, and international corporations, to the catholic (universal) Church. In every form of society, management is constantly struggling to harmonize individuals' particular ends with some common goal, the reason for the group's existence (*Mitsein*). Social forms are desire-driven and necessarily reasonably harmonic in their dynamism, otherwise they fly apart.

What I am going to point out next may sound rather illiberal. If by liberalism, one means a vision of the world in which every individual is free to do whatever he wants, then, yes, social reality is essential illiberal. But I am sure you will agree that extreme individualism is not genuine liberalism but *libertinage,* much practiced, rarely defended theoretically. The achievement of practical truth, the common end, always requires some sacrificing of individual desires and subordination of all the remaining tolerable or needed desires to the common goal; hence no society, familial, corporate, national, or ecclesial-universal can be maintained without discipline. The disciple is guided by the leader's vision. The goal cannot be reached without order, which has to be lived out knowingly and concertedly. Truth is, as Hans Urs von Balthasar says, *symphonisch,*—and a symphony is "harmonic" as it unfolds in time, achieving the composer's goal, creation of a completed dynamic form, an unfolded world of sound.[1] (Dissonance is introduced by a composer very deliberately, against the background of the prevailing harmony. Dissonance arising spontaneously in a social form requires the conductor to call his musicians into line.)

Institutional structures are devised to assure this dynamic harmony, and individuals are acculturated to these traditions of practical wisdom—"how we do things here," (just as musicians learn to play their instruments and how to follow the score). The newly born or the newly inducted in any institution are in-formed, both by imitating those they see around them (thus *implicit* tradition is passed on) and by being told (this requires language symbols) or motivated (through liturgical, artistic and scientific symbols—and propaganda and advertising). As he is inculturated into various contemporary institutions, the young person absorbs a lot of *HTX mentality,* forming mind-sets harmonious with his peers with whom he must cooperate. (The revolutionary, like the composer introducing dissonance, deliberately exacerbates existing disharmony, but in this case to destroy this living truth, or as he sees, this conspiracy of lies.)

But can the HTX be said to have a "form"? Does enough of a global harmony yet exist to manifest something definite, something like a common project? Is anyone trying to identify its ends and see to the well-being of the overall HTX? Who is seeing to its order? Where is the conductor? In certain large sectors, yes, there is concern for the well-being of the planet: think of the International Monetary Fund, all the UN agencies, the environmental organizations, and, indeed, the Church, concerned (as it has been since Christ) for the unity of mankind. These large institutions have their directors and their popes. The HTX is certainly not endangered by lack of structure, like a crowd in a burning theater rushing to the exit, only to produce a deadly crush. Every effort to manage large-scale processes in the HTX in ways informed by HTX mind-sets expressed in HTX symbols is a contribution to maintaining and developing large elements of *what might be emerging* as the englobing form of the HTX. The nascent form is emerging from the anatomy of its vast institutions, through what cooperation there is

---

1. Hans Urs von Balthasar, *Die Wahrheit ist symphonisch. Aspekte des christlichen Pluralismus* (Einsiedeln: Johannes Verlag, 1972).

among them, and in scientific, technological, political, and artistic symbolic forms.

Still, the Internet is remarkable precisely for its chaos, a certain formlessness, an ordered disorder. It can provide us with a flood of data on nearly any subject we care to name. But being overwhelmed with data does not necessarily lead to our being better informed. The question remains: Does the being of the HTX itself reveal a clear enough set of ends to allow constitution of a form, or at least a massive, deliberate striving to produce together a planetary social form, or are we seeing a ragged interplay of vast forms, dialectically affecting one another in a formless openness?

We need a more explicit account of what the HTX being may be calling for in the way of institutional arrangements, some of which are emerging before our very eyes. (Recently, the fiftieth anniversary of the founding of the United Nations provoked much reflection on how to reform it to achieve "more efficiently" its planetary-scale duties. As the thought of "world government" frightens, with power centralized in the hands of a few at the top of a vast bureaucracy, I personally hope the HTX may encourage a looser zillion institutional interactions that will stay loose. As bigger structures appear, for example, the European Union, we see an increasingly articulate group of world observers alerting us to the crushing nature of big government and faceless bureaucracies.)

Existing social arrangements for controlling and molding relevant natural and social *processes* appear in several kinds of form.[2] These types are often distinct in function of scale. For instance: if the group is small enough and all members understand (at least implicitly) what is to be achieved, the form may be that of a mutually beneficial, perhaps even loving, cooperation among equals. A larger group may take the form of a dictatorship, perhaps called a "democracy," ruled by leaders with a cynical, will-to-power manipulation for control, or of a loving (or brutal) patriarchal hierarchy.[3] We must watch for new forms rising from the HTX complexity and the planetary extent of new social groupings, and try to understand the pressures that mold them.

To sum up: Because there is in it no overall authority structure, the HTX is not strictly speaking "a social *organization*." *It is rather a loose,*

---

2. Recall that processes are sustained courses of change consistent and enduring enough to have come to the attention of one reflecting on what is happening in a social situation or unfolding tradition.

3. "Paternalistic" is the best label for this kind of authority structure, which is patriarchal whether the chief is a man or a woman.

even ragged, planetary, singularly complex set of interactions ("the Web")
between many kinds of social organizations at all scales, each with its own
goals, yet many such organizations cooperating and competing within
larger structures, all englobed within the illumining being of "the HTX
world."

That being (*Sein*)[4] brings into play a group of characteristic, mutually
reinforcing mind-sets expressed in new HTX forms of symbols and uti-
lizing a vast array of HTX hardware. But it has not as yet led to the creation of an all-englobing control-exercising institution with a focused
common vision. And maybe it never will. And perhaps that is a good
thing! There may not as yet even be *an implicit order* for the realization
of which a responsible leadership should be molding the cooperative
actions of billions. Talk of a "New World Order" may be misleading.
The new HTX stresses openness and multiculturalism. It also fosters individualism while encouraging giant organizations, setting up a vast
tension between egocentric aspirations of the little individual worlds
and the bureaucratic demands of the large organizations. Seeing those
inherent tensions, one cannot now reasonably predict whether the HTX
being will lead to a crushing centralization (Remember Tiananmen
Square, and Teilhard de Chardin's nightmare of "*le grand Molloch*"!) or
anarchy fueled by increasingly dysfunctional and socially inept egoists,
or (more hopefully) just maintain the current tense balance without
breaking down into world military strife or widely disrupting nihilistic
violence.

This planetary-scale, partially disorganized, emerging social being
confronts us then with a problematic reality. For, like it or not, fully
formed or not, the HTX *is;* the widely propagated mind-sets we shall
examine *are* illumining the vast but loose weave of institutional interactions, many of them very new, very HTX in character, very similar in
style. But, as I said, they remain without any clearly focused goal beyond a general pursuit of profit, efficiency, good management of the
institutions' own affairs, and a (dangerous) will-to-power, more remi-
niscent of a crowd rushing to the theater door than of a well-ordered
family. All of this is creating and distributing heretofore unimaginable
wealth, providing new forms of education, inducing explosive (but

4. I suggested earlier using the German word to remind us of Heidegger's development of this sense of being as the englobing illumining that makes possible
"world." In *Being and Truth,* I distinguish this more subjective *Sein* from the in-itself
reality of things and persons, which I symbolized by *esse*—their act of existence,
their to be, and I explained how both Sein and esse are present within "the analogy of being." See *Being and Truth,* chaps. 9 and 10.

now slowing) population growth, polluting the earth, and covering the landscape with real networks and architecture we have no difficulty recognizing as at least industrial, if not yet HTX.

Some Ragged Interpretations in Search of a Ragged Form

At the start I could see no other way of getting a sense of the HTX's peculiar being than, first, seeking to identify its *constitutive elements,* then, second, discovering how they interact. It is very HTX to use acronyms (Dilbert and Wally invent acronyms that stand for nothing, just to flummox their pointy-haired boss). Here goes: The HTXRG has been able to identify about eighteen "software" factors and six big groups (*genera*) of "hardware," each breaking down into myriads of new kinds of instruments.[5] To satisfy your curiosity I shall list them, afterwards explaining some misgivings that overtook us after this research.

SOFTWARE
—ideas (product, plan, what do you want to do and control? doing rather than being; designing for easy manufacture and for marketing)
—thought of the whole, temporally (eschatological, progress) and spatially (global and gigantic)
—centralization-decentralization (in bureaucracy)
—management, manufacturing (including automation), transportation and marketing techniques
—creativity vs. imitation (tension between future and the already having been, between loyal conservatisms and radical utopianism)
—efficiency—by what criteria (often absurdly narrow) inculturates its own way, and then destroys its inculturations, lest they resist totalization (hence bureaucracy's lack of loyalty to employees undermines commitment on their part)
—investment capital (store of value plus putting resources to work
—monetary system (everything mediated through it)
—"everything for sale" (culture, entertainment, sports)
—personnel (staff, workers, customers, citizens)
—mobility (interethnic)
—education vs. manipulating public opinion

5. HTXRG is the research group of cutting-edge techies who worked together in preparing this study.

—population explosion
—family and personal breakdown
—social welfare net
—materials and energy
—concentrations
—information (a resource, needed about all of the above)
HARDWARE
Particle Controllers:
—electricity generators, transmission lines, storage
—x-ray machines, radio frequency transmitters, microwave devices, nuclear medicine machines
—devices harnessing these capabilities for communications (telephone, radio, and television)
—computers
Molecular and Cellular Manipulators:
—vaccines, antibiotics, genetic engineering devices
—fertilizers, pesticides, improved seeds
Optical:
—eyeglasses
—micro- and telescopes
Other Power Sources:
—propulsion devices (steam and internal combustion engines, jet propulsion, ram and rocket engines)
Precision Automatic Tools and Robotization
Newly Exploited and Artificial Materials
—stronger, lightweight metals
—plastics
—new textiles
—synthetic rubber, etc.

Once we had assembled this formidable list, based on rather old-fashioned abstractive thinking, it began to look to us, in the light of the new HTX, like a flawed approach to the problem of essential description. Bruce Stewart more recently suggested it is probably better to inquire in terms of "motifs" and "constructions of reality" than to point to piece-parts. We shall get a hint of these motifs through certain HTX symbols we found popping out in discussion of those twenty-four abstracted factors. Reflecting further on these motifs led to glimpses of certain all-englobing notions—what I believe Stewart meant by "constructions of reality"—underlying the whole vast reality. These glimpses bring us close to its essence, as we discern its peculiar proto—

or emerging—ragged and fast evolving *dynamic form,* with its old (industrial) and new (web) Gestalts in tension.[6]

So we looked over those essential dimensions and some of their interactions allowing those key symbols to jump out, as we struggled to find some term for each identified dimension.

Ordinary language is the most dominant kind of symbol system in our lives. I am using the murky word *symbol* here in the widest general sense, for any sign, gesture, or artistic expression, whatever its emotional charge. The ability of a symbol to move us depends on the way it is used and by whom and in what circumstances it is received. A stop sign gives a defined instruction. A metaphorical use—"The road to personal progress is strewn with stop signs"—is clear because the word's basic signal is clear, but the emotional charge is raised. When a young woman tells a young man, "When I see you, a stop sign pops up in my face," the emotional charge is probably quite high for both speaker and listener. But in addition to ordinary language we use many less ordinary languages, like those of the law, ritual, full-fledged liturgy, art, and mathematics. All play large roles in the HTX. (So does "technobabble," an ordinary language for the players in a narrow HTX sector, an arcane ritualistic language for those outside it. The *Economist* once imprudently estimated the number of technical words in English at ten million. That is just a way of saying there are a lot!)

Man is not just a practical animal, he does not group only to produce, reproduce, and survive: he can also contemplate and appreciate, as well as dissipate, grouping for a party just to have a good time and "to blow the mind," (quite a symbol, that!). Most of us muddle along in life, just trying to survive, distracted, juggling many balls, interpreting as best we can many different kinds of symbols.

Indulging in distracting fun is one of the dominant aims in wealthy HTX societies, and much of art and entertainment is simply *play* with symbols. Many entrepreneurs and corporate climbers and politicians are motivated in part by the fun of the game. And how much of Internet surfing—a pure symbol manipulation—is not *serious* at all? Entertainment (including tourism) is the largest industry; the new 110,000–ton "Love Boat" is the largest ship, a floating symbol of escape, exceeded only by the hypertankers bringing the lifeblood of industrial society.[7] Unfortunately, entertainment too often is no longer a

---

6. *Spüren des Seins,* glimpses of Being, we might say, "motifs."

7. Oil is to the industrial core what cash flow is to the whole HTX. Big films produce fabulous cash flows but no fuel!

recreational distraction; it has come to occupy the head and heart. Surfing the net may be the new poor man's tourism or an electronic junkies' trip. Symbols can be possessive. The "truly successful" however get their kicks from "playing the power game," with those characteristic symbols—big titles and big stock options—as measures of success. I shall leave for another work, a philosophical anthropology of postmodern man, any further reflection on the proper place of play in healthy human existence. In heaven, the angels sing (a symbolic art form), but Christians believe we get there only through the Cross, which is no mere symbol and decidedly *not fun*.

Symbols are deadly serious business. It is through, and in terms of, his symbols that man works out his conceptions of who he is, what he is doing, and how his world is constituted. This has been true since the earliest myths. How much of peoples' notions in the HTX of what they are, and are expected to be, come to them through television drama, talk shows, and film, streams of symbols weaving new mythologies?[8]

There is no overarching language of the HTX, not even bad English, Boolean algebra, or post-rock music, widespread though they may be. Yet the being of the HTX is transforming old languages, creating floods of lingo and technobabble, and aiding the generation of new art forms and strange liturgies. All this is happening with no overall coordination, yet *with an uncanny similarity of style*.

The nihilistic tendencies of this style become clearer by the day. The mysterious growth of languages is a fundamental phenomenon of being's coming to be. It is never simply the fruit of plots to manipulate, order, and control, not even when the court of Louis XV controls "L'Académie française," the *Reichspropagandaministerium* controls education and the media, or political correctness patrols work deliberately to spread hyper–self-consciousness through a Fascistlike manipulation of ordinary language.[9]

8. I was revising this text on the day of the Québec referendum, which is a step in deciding the fate of a nation. The campaign was a war of symbols, with the Québec sovereigntists clear masters of the rhetorical art, the Federalists muddying the waters with facts about the economy and the difficulties of rearranging hundreds of treaties. Many "soft" separatists would be mollified by a flourish of rhetoric: language in the constitution acknowledging Québec as "a distinct society." The pragmatic and resentful WASPS want to be sure that does not accord Québec any advantage. The final vote—a 51–49 split—leaves Canada unstable.

9. Here is an example of resolutely non-nihilistic use of linguistic symbol: Since the days of the fathers of the Church, the bishops have taken utmost care to

A separate monograph called "HTX Art and Entertainment" should be written to probe the reasons for the attack on beauty, the seemingly deliberate cultivation of ambiguity, and the nature of many of the manipulations involved in HTX symbol abuse ("cultural rape" Postman calls it) at its worst.[10] In order to *move* people to act according to plan, symbols are used, not just as orders, but as suggestions and subtle, even manipulative, motivations: Enter advertising, propaganda, and politically correct language games. From the beginnings of spoken language, the articulate have convinced others to act through their adept manipulation of symbols. These linguistic possibilities can even be used for entertainment and appreciation; all motivating through symbols does not have to be sinister manipulation. (These last depend on playing with the emotive capability of gesture and symbol of every kind—including mathematics [Steven Weinberg, the Nobel Prize–winning particle physicist, is eloquent on how the beauty of a mathematical demonstration motivates its acceptance, and he obviously prefers to spend much of his life in pursuit of that kind of beauty.][11] The emotive evocation of whole levels of being is an essential role of art and liturgy. The two can also create a "virtual reality" most useful for escaping harder kinds of reality; the arts as well as religion can be opiates of the people.)

Fortunately, however, being is never simply *controlled,* not even the being of the smallest social units, like the family, not even the being and destiny of the free individual, certainly not controlled by his own will alone! Even in religion and art objective reality can show through the medium of the shaman's or the artist's constructions. The true

---

protect the founding symbols of Christianity, many of them direct gifts of Jesus Christ himself, and of the prophets of old, and to strive to see that the languages into which they are translated are adequate to the being they express. At times, an anti-Christian spirit has crept into what was intended to be a fostering of the truth, at the worst of times becoming thought control. But the Church is always learning and today continues to constitute the profoundest concerted worldwide discussion of the sense of symbols. The confusion of symbols in civil law, due to lack of a good ontological base, stands in stark contrast to this. The nihilism resulting from increasing manipulation, instead of loving receptivity of being's gifts of deep symbol, is frightening.

10. E. Michael Jones, *Dionysos Rising: The Birth of Cultural Revolution out of the Spirit of Music,* traces the connection between the sexual mores of key figures, from Wagner and Nietzsche to Mick Jagger, and the ways in which their music has affected HTX society (San Francisco: Ignatius Press, 1994).

11. See "Beautiful Theories," in *Dreams of a Final Theory,* cited in Alan H. Guth, *The Inflationary Universe* (Reading: Addison-Wesley, 1992), 219.

prophet and the true artist forge and use symbols in the service of the objective and the transcendent.[12]

Symbols do not float in an ethereal idea space into which we reach up when we wish to represent something. *Au contraire,* within a given "little world" (*cosmion,* plural *cosmiota*), there is a dialectic between ends and means, between things and symbols, and, for that matter, between symbols and symbols, which form complex systems of meaning, influencing our actions. The *being* of the HTX, by enveloping myriads of *cosmiota* and their often tenuous interrelations, with all their peculiar languages, illumines an unimaginable stew of interpenetrating symbols. The great symbols expressive of the all-englobing notions that yield glimpses of the HTX being are themselves given substance by the little worlds that are its building blocks.

Now you see why it is important to reflect on the deeper sense of the symbols the HTXRG has rather spontaneously chosen to represent the various abstract dimensions of the HTX, in search of the most all-englobing notions.

The Greatest Social Phenomenon of Them All: The HTX

So from a modest incursion into the early industrial revolution we make an immodest effort (philosophers are not paid to be modest) to step back behind the being of our epoch, to glimpse as best we can from impressions of the interaction of its various dimensions, *the unfolding essence* of the largest social reality that has ever existed, to the extent some form is emerging, in its most recent, ragged gestalt, "the new HTX." We start with descriptions of various real elements, but we shall be attentive to the symbols through which they are expressed, their sense, and the broader significance embedded in them.

Scale, Population, and Physiognomy: The Spread of Civilizations

We begin with a reflection on scale and demographics, because for incarnate space-time creatures like human beings, scale is vital, and nothing is so important, nor more *real,* than the mass of 6 billion human centers of awareness and initiative, how they are grouped, and how their actions are molded.

Heretofore, the greatest social structures—civilizations—spread from a single homeland (often a single city) to contiguous territory, either by military conquest or a combination of alliances and conquest.

12. On truth in art see my *Tradition and Authenticity,* 69–72, 101–2.

The Church permeated the Roman-Hellenistic civilization and moved out to the wider world with the conquistadors. The institutional backbone of civilizations was always well focused, usually in some form of imperial state.[13] The Church has a unique structure of its own, monarchical and hierarchic. Islam, which spread from Medina and Mecca, has no very controlling overarching institution, even the caliphate having been quickly weakened.

Modern European civilization, with the rise of national states, broke with the ancient pattern of expansion. With the exception of Russia, which moved from a core out in the old way, modern colonial movement began from *multiple centers* within Europe and, leaping the seas, opened up a new scale. This reach was made possible by technological advances in maritime navigation and powder-fueled firepower and *by a change in mentality:* a powerful mix of Catholic evangelizing, reaching out to all men (followed by fleeing Protestant sectarians—Puritans, Mennonites, Huguenots—seeking new lands to build the Kingdom of God and manifesting in a new competition, born of a beginning disaggregation of the medieval Catholic civilization), and a new national consciousness, part of that same disaggregation of international Europe: So now it was "Conquer for the glory of Spain!" or to create in the wilderness a "Novo Ordo Seclorum," as it still says on the U.S. dollar, in a conscious break with the Old Europe.[14] These were all new symbols! Not just *libido dominandi* but powerfully moving visions captured in symbol changed the face of the world. The world had never seen so radical a revolution, not even the spread of Islam compares. Welcome to modernity, itself a new concept represented by a new symbol.

The importance of the European imperial civilization's technical superiority in navigation, in arms ("the gunpowder revolution"), and in its sense of political organization for forging modernity can scarcely be exaggerated. Both rapid technical advance and political strategy were the fruits of newly developing mind-sets, new capacity of conceptualization and imagination. What was striking about these

13. Related to this is the question of the relative importance of military power in the HTX today. I shall speak to that in the conclusion.

14. Why did Islam, which is also missionary, not spread far and wide by sea, once technology had advanced? My hypothesis: the Muslim countries lagged technologically by the sixteenth century, and hence could not mount a successful sea attack on distant outposts. After Lepanto, Muslim control of the Mediterranean was lost. They did however sail east, resulting in the Islamization of much of Indonesia.

technical advances was their exponential leap in range: the firearm meant one could attack from a distance with more power, and the canon could batter down the bastions of the feudal lords; the advantage had shifted from defense to offense.[15] There was a corresponding, no less startling, leap in the range of political imagination: Cortez was able, with four hundred armed men, to defeat the Aztec empire by striking alliances with the oppressed Indian tribes; De Soto and Coronado, on the other hand, got nowhere with their expeditions to the weaker tribes of the American Southeast and Southwest because they were interested in plunder and had no sound plan of settlement. It is harder to understand the leaps in political and strategic thinking, though they are well documented, but obviously something changed fundamentally in the mentality of the Europeans as these peoples (some of whom, like the Dutch and Portuguese, were hardly numerous) came to control vast continents.

This worldwide adventure laid the foundation for a European, high technology, increasingly urban international civilization, which eventually, as "modernization" made its inroads into all but a few peoples, planted "nodes" for the future HTX in the heartlands of all but a few cultures. The gradual internal transformation of that once-Christian civilization brought into being the contemporary "X," which itself is no longer a civilization, indeed it enfolds hybrid "modernizing" civilizations within it, nor does it now belong exclusively to any traditional culture, however much it has absorbed of various European and American cultures (and now elements of Eastern religions in New Age concoctions.)[16] It is spinning new subcultures of its own (e.g., new technological traditions, new traditions of sports, generating thousands of cosmiota of an utterly new kind, and think of all the other bizarre forms of entertainment worlds, including Disney worlds, Ole Man River worlds, and "Rock culture" worlds.)

I agree with the author of a recent article that "culture is an imprecise and changeable phenomenon that explains less than most people

15. I believe James Davidson and Lord Rees-Mogg are right in contending in *The Great Reckoning* (New York: Simon and Schuster, 1993) that a similar radical shift of advantage to the offense has occurred in our day, with the terrorist now being able to pack in stealthily enormous disruptive explosive power; witness Oklahoma City. They are also right that this is another element in the descent of power to the lower level, hence decentralization is in.

16. Some would call the new HTX postmodern. This is a term I avoid, because of the possible false signal that much that was modern has been left behind; rather much of Enlightenment mind-sets is being criticized with new vigor, but so much

realize."[17] But you can say that for most of what is essential to human existence. The first primitive bands of humans formed cultures. Civilizations built on previous cultures, transformed them and developed "high cultures." The HTX builds on many cultures, basic and high, coming from Christian civilization, but it transforms them, while producing many new cultures of its own.

I understand a culture to be webs of capability founded in and built up through previous accomplishments of human transcendence, traditions expressed in symbol systems and handed down through education, stored in memory and habits of individuals, and engraved in cultural objects (books, buildings, artworks, liturgies). Much of the passing down of cultures is implicit-imitative, but more explicit education gains in importance as the cultures become "higher."

The bizarre forms of entertainment worlds, Disney worlds, Ole Man River worlds, and "Rock culture" worlds, each have their own "traditions" and symbol systems, but they are not always true *cultures*. They lack deep roots. Some are completely unrooted, belonging in the soil of no place, not wedded to conditions of land and climate, they are even poorly rooted in human nature, being often more manipulations of superficial emotions than rich forming of the depths of human potential. The difference between genuine HTX culture (for instance, scientifically based technological transformation of atoms, molecules, and genes) and the anticultures that uproot more than they settle merits more thought. The pseudocultures are but poorly anchored social forms of new kinds, requiring more ontological investigation and a new generic name ("HTX scams"? "Profit centers"? "Cults"? From this rhetoric, you might guess I don't think much of many of them!)

------

of the Enlightenment mentality remains imbedded in the HTX mind-sets. Recently Bruce Stewart muttered the heretical thought, "Maybe there is no HTX, it is just Western civilization working itself out." The HTX originated in, and continues to be dominated by, Western civilization. However, the instantaneity of the multi-nodal character, the multiplication of these nodes by the hundreds of thousands, and the fact that while populations continue to congregate in the great cities and a certain number of HTX capitals wield immense influence, the HTX is not essentially city bound, nor centered really anywhere, and increasingly electric, are signs that the HTXRG has been right to search for all the characteristics of HTX that transcend Western civilization while not forgetting how modern, Western, and even Christian it remains.

17. *Economist*, Nov. 6, 1996, 23–26, with a bibliography of recent works on culture and its influence.

The HTX Spreads from Nodes

Penetration from Node to Hinterland

After independence, the nodes of modernity in the former colonies continue to work their transformation as they penetrate the hinterlands, establishing ever more satellite subnodes.[18] This process is not unidirectional. Modernization goes out from the center, but migrating rural populations flood to the outskirts of the modern city, producing strange dynamics. For instance, the fruits of modernization penetrate the *barrios d'invasion* (as Colombians call the squatter suburbs on stolen land) unevenly, the poverty and unrequited drives of the recent migrants pressing dangerously on the resources of the more wealthy urbanized core. In the American inner cities, recent immigrants from poor Latino rural areas mix dangerously with an urban underclass formed largely of that minority of African Americans who have not assimilated well into the HTX life of the city.[19] Meanwhile the middle class (including economically successful blacks) withdraw to the suburbs or into well-guarded condos. The world-scale HTX has its more widespread "inner proletariat"—all the Fourth World peoples it may never transform, but who are surrounded by the HTX networks, which they in many ways infect and can disrupt. (Ask the folks in the World Trade Center about that!) In China, India, Indonesia, and Africa, which retain huge hinterlands only minimally penetrated, the two-way flow from town to country and back continues to cause severe disequilibrium: unassimilated peasants in the city, returning city "dudes" contaminating village life with half-digested city ideas, forming local élites who can, as in Africa's Lake Country, use HTX firepower to create genocides as they maneuver politically.

Perhaps as much as half of humanity remains only superficially affected by the HTX. Two billion "centers of initiative," barely transformed, constitute a large reserve for mischief as they move more into *the* HTX world. Is the HTX network of whole nations and strong HTX sectors in many Third World countries strong enough to resist the convulsions that can upset the Chinas and Indias of this world (and so much

18. For example, Bela Horizonte and Brasilia could be seen as satellites of Sao Paulo and Rio de Janeiro.

19. Development has so penetrated the countryside in most of North America, the industrialized farm territories, that only nuances of difference in lifestyles and mentality separate the HTX farmer from the merchant in the city. There are exceptions: Appalachia, Indian reservations, and, most significant, because so close to the heart of the HTX nodes, the "inner city wild west," a startling Toynbeean "inner proletariat" phenomenon.

of Islam), the pressures of North Africans on France, Yugoslavs and Turks on Germany, Indians on England, and Mexicans on the U.S.? Toynbee saw "the internal stability produced by the unassimilated, furnishing that "inner proletariat," to be the perennial downfall of civilizations, and more and more of these peoples are forming internal proletariats in highly HTXed countries. A complacent and passive large population in the rich heartland of the HTX could spell danger when serious troubles, interior and exterior, have to be addressed.

The HTX nodes remain tied by all the bonds we shall describe to the multiple world centers of the HTX: Bogota more to New York, Nairobi and Madras to London, Taipei to Tokyo and Hong Kong. Along these conduits the HTX is transmitting distinctive cultural forms from the deeper levels of the old transforming civilizations; also the Third World nodes are sending back through the net some of their deeply rooted wisdom: Imagine, microwaves transmitting the Christian gospel, the Muslim call to submission, and Confucian family influences! (What does it do to these symbols to be zapped around the world at the speed of light and then half-assimilated by people with half-baked understandings, not formed by long contact with the underlying cultures?)

Meanwhile between these world centers themselves an ever-denser net of relationships has been woven. Each node (city state, megalopolis [e.g., San Francisco–San Diego], or even continent-wide network [the HTX network of U.S.–Canada and of Western Europe are becoming like giant planetary-scale supernodes]) constitutes a center of intensity from which "radiates" the HTX, first to satellite subnodes (regional centers, necessary for distribution of goods and services, and special concentrations of manufacture, mining, forestry; the recently emerged powerful nodes of "the Asian Tigers," for instance, radiate to their hinterlands, but also serve as a kind of secondary center for the supercenters); and then more diffusely out into the countryside. The whole "planetary hinterland" has become the field of play for the North American, Western European, and Japanese supercenters in competition, some of their influence mediated by the Tigers and their Latin American less brilliant counterparts. Tensions fed by the resulting cultural conflicts are increasing. (The Persian Gulf situation is an example worth considering, as are the turmoil in Hong Kong, Singapore, and Taiwan. The recent serious problems of Asia show the fragility of many of these relationships, and the difficulties of evolving quickly enough institutional structures, such as an adequate international monetary system. All of these transforming nodes are still profoundly affected by their respective ancient cultures spawned by great civilizations that resist at the core. "The clash of civilizations" is today really a set of tensions within the powerful overarching HTX web. Civilizations have clashed from the beginning. This is different, as each of the civilizations is being transformed in its own way by a larger noncivilization).

The New Movement: The Weaving of the Internet

One critical counterfeature to this spread-from-the-node model, we have already suggested, is the Internet, the existence of which alerts us to the fact that we have entered a dramatic new stage of HTX development: the HTX phase two. The military developed the original Internet specifically *to get rid of central nodes,* because if a few were to be shut down by nuclear attack, that could take out the entire network. The process of decentralizing so that there is not one node but, rather, hundreds of thousands of potential nodes is fundamental to the *form* into which the new HTX is now evolved. For example, phase one HTX: a large New York bank has a complete replica of its head office built and maintained at a second location so that in the event of a natural disaster the "node" does not go down. The bank merely switches over to the backup. Phase two: In the Internet, any one location can become a central node, depending on the evolutionary requirements of information transfer. If less mail is going to New York, the routers in that area will lose their central status as other nodes become more important. Furthermore, if one node goes down, the system is designed automatically to find a suitable path so that information reaches its destination.

The need to expand markets and secure natural resources and the need for new technology is as old as the industrial phase. The expansion of the variety and improvement in quality of services is phase one HTX. The acquired experience and much developed mind-sets of phase two, as well as the improved communications and reliable jet travel, intensify the willingness to diversify enormously the net of commercial and technical interactions. Even as recently as phase one it would not have been conceivable to build jetliners with synchronized construction of major components in five countries. And it was difficult to survey distant nodes by telegraph only. Many of today's world-traveling HTXers would be reluctant if they had to travel by steamer.

Untransformed cultures from earlier civilizations still prove both a drag and a boon. Recently the small high-tech company in Ontario of which I am a part owner won an ongoing contract with Northern

Telecom to supply harnesses (complex wire assemblies) even though we bid $100,000 more on the first million dollar order than the company in Guadalajara that had been assembling them, and we can be very profitable while paying $7 more an hour to our (largely Vietnamese and Indian) employees than workers earn in Mexico. Our advantage? Reliability (utterly essential for HTX manufacturing) due to better management (not just HTX but humane!) and the attention our workers pay during monotonous operations, because they are devoted to the company. Nortel also cited "ease of communication." When a problem arises, we are only one hundred miles away and hands-on solutions can be quickly found. The old cultural reality of face-to-face cooperation remains.

This tightening weave bringing many new centers of initiative into existence produces a ragged world social network—worlds within worlds, worlds leaping over old worlds to connect with new—of a scale, complexity, and dynamism *unimaginable* even two centuries ago, and growing in quantum leaps from HTX phase one to phase two. Now the new open net of communications, with information being offered for free on the Internet, with new web pages and bulletin boards and files pouring forth a cornucopia of data for the asking, allows skipping from here to there to anywhere at the speed of light. Upon these purely HTX relationships, strong resistant natural and traditional relationships exert a drag. These are furnished not only by raw nature itself, and language and residual cultures, but also by those traditions kept alive by the evangelizing world religions, Christianity and Islam, the vibrant force of which can be transmitted by HTX communications. (Yes, the Church, as we speak, is in fact strongly present on the Net, and Muslim associations have web sites.) Through these religions the HTX lands are further tied to a foreign center, the mother culture or the home center of the religion, while tensions build up in cross-cultural clashes across vast oceans.[20]

Nature at the Base: Demographics and Ecology

Nature constantly reminds us, however vast and multifaceted the weave of these human arrangements, that she is always there at the base, mostly unreformed. Through the engineering of new kinds of molecules we may be able to produce undreamed-of materials, and

20. Brazil to Portugal and Rome, Argentina to Spain and Rome, Indonesia to Mecca and Cairo, and New York and Manila to Rome, Jerusalem, all of Italy, and Dublin (remember the St. Patrick's Day Parade, Me Lad!) and London!

engineering genes can produce basic changes in human beings, but the 98 percent of the cosmos consisting of hydrogen and helium atoms is largely beyond our control, the great plates of our planet continue to float on the ancient, cooling magma, pushing up Everests and Denalis, erupting into Rainiers and Etnas at will, and the weather remains unpredictable. And don't forget the mystery of the cell.

So it would be a good idea to found our study of the essential elements of the HTX on a respectful consideration of Mother Nature, not that *factice* revival of an old pagan symbol, HTX de-formed, Gaia, but that nature which does bring us forth indeed.[21] Mother Nature is a symbol for the reality of our biophysiological makeup, with roots down to the particle level, and in formation since the Big Bang. Because questions of *demographics* and *scale* are so fundamental to human affairs, and demographics are especially important to the task of motivating and controlling masses of people to achieve social order, let us begin with an effort of the imagination trying to get some hold on the reality of 6 billion nodal centers of awareness and initiative increasingly networked.

High-tech medicine and industrialized agriculture have worked their magic on nature, making it possible for this planet to bring forth millions at an exponential rate of increase and sustain them, with increasing life expectancy. Since the time when 1.5 billion lived with a life expectancy of, say, 43 years until now, 6 billion with a life expectancy of perhaps about 64 years, the planet's carrying capacity has expanded 550 percent. This explosion altered the human situation fundamentally; it has happened with overwhelming suddenness. It is now calming: 86 countries have birth rates of 20 or fewer births per 1,000 population (21 is replacement rate; Canada's present rate is 17); countries with high birth rates represent only 21 percent of the world's population, while those with low rates represent 45 percent, according to the U.S. Census Bureau. Fertility rates in 59 countries are below replacement. China is now one of them. Nine countries—all parts of the former Soviet empire, suffering from that regime's anti-life effects—actually have declining populations. Populations from the Third and Fourth Worlds are increasing in proportion to the world's total, with rates decreasing as they develop. The rate of development and rate of

21. See James Lovelock, *The Ages of Gaia: A Biography of Our Living Earth* (New York: W. W. Norton, 1988), for a New Age myth of the goddess Earth who quasi-intelligently maintains a living harmony. The complexity of the necessary harmonies is beautifully invoked.

population increase have proven almost exactly inversely propor-
tional. We have yet to recognize the radical impact of the new scale of
our populations—a Mexico City with 23 million, (compared to
Imperial Rome's 4 million at its apogee), a China with 1.2 billion—on
all our systems, and through them on our daily lives. Why are the implications of this fundamental fact so hard for us to grasp?

Enter a basic fact about commonsense human reality: Day-to-day
psychological survival requires us to take just about everything for
granted. This taking for granted of the most marvelous and mysterious
(and ultimately significant) realities is a form of normal defensive
mindlessness, a necessity of short-range pragmatic action. "Doing"
and "contemplating" are hard to pull off at the same time.

If half of this huge population has been seriously acculturated into
the HTX, that constitutes a social reality touching the lives of 3 billion
people in some 180 countries.[22] Acculturation into any social reality is,
of course, by degree. Such factors as social class, distance from the
center, and adaptability of peoples of various cultures to the new HTX
play a role.[23] The some 700 million peasants of India, the 850 million
of China, the 100 million of Indonesia, the 40 million of Egypt, and the
70 million of Pakistan constitute a third of humanity as yet little influenced by the HTX. Add to this the peasantry of Africa, the Philippine
Islands, and the highlands of South America, and one arrives at close
to half. But then think of the fact that 90 percent of the engineers who
ever existed are active now. You begin to see how the clustering of millions of engineering firms, industrial plants, and infrastructure projects
creates perforce an intensity and vastness of interaction beyond anything earlier centuries could have imagined. The necessary springing
up of fast transport and lightning-speed communications generates a
paradoxical movement, and strange "distances"—*the very sense of
space changes,* as weird new spatialities, like "cyberspace" and "virtual
reality" are born: On the one hand, the HTX spreads from the multiple
first centers to the other side of the globe, 10,000 miles or so away as
more and more hundreds of millions are caught in the web of the

22. And through HTX medicine and food aid most of the rest has been affected
to some small degree.
23. The HTXRG has developed a matrix of factors with measurable criteria to
allow some rough measure of various regions'/peoples' absorption into the HTX.
Indices of degrees of industrialization (of which GNP and electricity consumption
per capita are not bad), serve as a rough measure of degrees of HTX penetration,
account being taken of flukes in the case of the resource monsters like Saudi Arabia.

emerging world system; and on the other, the existential distance narrows, so that certain Austrians have become my close partners in operations in that country, with contacts much more intimate than with any of the neighbors living on the same street for thirty years.[24] Close bonds tie Wall Street and The City of London, but a chasm separates the towers of Lower Manhattan from the ghetto of the South Bronx.

The pressures of these demographic realities and the tensions of the new space-time relationships will force leaders to become conscious of the HTX in a way that invites them

1. to form a new concept, "world populations";
2. to pose explicitly, for the first time ever, the questions of the practical implications of scale and existential distance;
3. to conceive explicitly of "economies of scale"; and
4. to raise the ecological question of the ability of the ecosystem to support such large masses using so much energy. (The spiritual movement from Malthus to *Second Spring* is enormous. Now we encounter the most sophisticated ideas of ultimate limits, e.g., the "thermal barrier": transforming energy at a rate that produces so much heat we risk raising the earth's temperature to levels dangerous to life.[25] At the present rate, one can predict a time when the earth's temperature will no longer support complex life forms—not so far in the future as you would like to think!

The HTX is the first society to be pushed to think through *on the political level* the (Judeo-Christian) questions, "Who is my neighbor? Who is the brother for whom I am keeper?" In the course of this, HTX thinkers analyze the HTX's anatomy, using symbols like *First World* and *Fourth World* to capture its concepts.

24. These nodes, someone has suggested, are "city states," replacing the nation states of eighteenth-century prominence. I am not convinced this is quite exact; there are some city states—Hong Kong and Singapore spring to mind. But it is more what Doxiadis termed *megalopolises* which should be considered, and also certain larger networks, such as the North American production machine; Japan, Inc.; and the EU. The present painful reorientation of the COMECON productive block shows how large and meaningful these continent-wide machines can become: 1) city states, 2) megalopolises, and 3) continental production units are all, then, real, effective socioeconomic units in the HTX. See later in the chapter the discussion of "blocs."

25. Every energy transformation produces heat; most of the heat released on earth remains trapped in the atmosphere; hence temperatures are rising—independently of the issue of the greenhouse effect. Even without emitted industrial gases trapping the heat, the atmosphere can only lose heat to space at a limited rate, and now we are purportedly producing heat at a rate superior to that.

The HTX "demographer," the "development expert," and the HTX "ecologist" embrace a vast sweep of time and space in their concepts, while striving to remain analytic, even finely so, and "managerial" in intent. Their new mind-sets have developed out of an earlier set of scientific and industrializing attitudes and projects that allowed concentration of energy on an unprecedented scale, in the pursuit of efficiency, as modern men rushed *to do* rather than ever being content just *to be*. In the quest for knowledge and control, people have developed an impressive array of kinds of techniques. Some are very recent developments of the being of the new HTX, but many of these new techniques are built on very ancient ones.

A technique is a certain kind of culture: it is a *tradition,* usually partially caught in symbols, for inculturating people into the know-how of a world, helping them to conceptualize what needs to be done and to acquire sets of habits for actually carrying out complex tasks.

The discovery of certain elementary laws of economics having to do with the production and distribution of wealth (*The Wealth of Nations,* as Adam Smith called it) and the spreading acculturation of these notions drive that foundational platform of the HTX, the industrial machine, even more than traditional desires for national glory.[26] Industrialization, as we saw, requires many new kinds of techniques. For instance it has generated a weave of *managerial techniques,* mobilizing and directing manpower; *educational techniques* to train manpower; *engineering discoveries* ("breakthroughs," events brought on by sustained searching—research techniques are involved), allowing the concentration and harnessing of natural energy; *fiscal techniques,* mobilizing and putting to most efficient use capital and securing needed manpower and raw materials; and *marketing and distribution techniques* to secure outlets and revenue. Each type generates its own symbols. (That is why English now contains the estimated ten million technical terms we mentioned earlier.)

While the tensions inherent in this expansion and development have long been considered those of a "capitalist conflict," we are beginning to understand that the ultimate *caput* is the social capital of all the needed techniques stored up in good minds.[27] But there is something

26. These are not, however, absent from the picture, as protection of living standards demands an aggressive expansion of turf to secure raw materials; manpower, both the cheapest and the more highly skilled; and markets.

27. Michael Novak, *The Spirit of Democratic Capitalism* (New York: Simon and Schuster, 1982) and other works, has developed this theme.

more basic yet than stored up know-how: brilliant inventive insight in furthering all of these techniques, whether in the form of the break-throughs measured in patents issued, or in fiscal inventiveness (after all, money has to be found to realize the fruits of the inventions), or in man-agerial prowess.[28] Social capital investment and cultural genius are, in the long run, even more important than the necessary accumulated capital stocks—buildings, machinery, inventory—which, as incarna-tions of yesterday's ideas, quickly become outdated. Development ex-perts today look at the huddled masses more as persons, who with acculturation (by training and educational techniques) become more useful to the world system, and less as cheap labor. In the new HTX the ultimate source of wealth is recognized to be the nimblest minds with the best access to information.

Techniques for Managing the Ecology

Now armed with the same, even wider sweep of vision, the same analytic techniques and "management systems mentality," here come the HTX demographers and ecologists to say, "Wait a minute! There are other costs involved in what you are doing, there are other goods to be considered than just the obvious unit costs of production." They set out to manipulate political systems to control the brute biophysiologi-cal giveness: populations are to be controlled, eugenics is in the air, and protection is to be assured for certain pristine states of the planet; species are to be preserved even at great expense. These environmen-tal experts are obsessed with preventing us from violating newly dis-covered "limits to growth." (There is a powerful HTX symbol for you!) Into the debate within the heartland of the HTX is being introduced (al-though not many really want to face it) deeper questions about growth: to what end? of what kind? how much? Not far below the surface one glimpses the ultimate question: What is human "Life" anyway? Creative ideas, fueling new products and new social arrangements are indeed important, but, say the HTX "limits to growth" types, we must ask *to what end* all this frenetic development?

The clumsiness of the self-appointed Club of Rome's pioneering effort (1975), *The Limits to Growth,* should not be allowed to destroy our sense of

---

28. Operation Desert Storm was as much an exercise in, and demonstration of, large-systems management skill as it was an invitation to certain obnoxious char-acters to respect lines in the sand. Russia took note, and the collapse of the Evil Empire was consequently furthered.

the legitimacy of the question they posed: Are there visible at this time limits to industrial growth, with concomitant energy-transformation and destruction of resources? Is "sustainable industrial growth" an illusion, and if so, what are the limits, and how should we manage the world system so as to avoid colliding with them?

The debate cannot help but turn on the more philosophical question: "What is man for?" What is the present generation's responsibility for generations yet unborn? How far forward should we cast our glance? And how do numbers of human beings—a matter over which we now in fact just may have some control—figure in all this? So many questions demanding something more than *techniques* to be answered!

The Club of Rome types are right when they insist that the energy transformation following on industrialization is straining natural systems. Concerned social psychologists are right to question the ever more frantic pace of life, as we all try to keep up with the inventive geniuses. Both groups are right in leading mankind to reflect on what is happening to itself demographically and structurally. But the raging debate about overpopulation, overdevelopment, and quality of life requires more attention to issues of WHAT does WHO want? And just Who is WHO? The control procedures most often proposed—and massively financed by wealthy individuals and wealthy governments, the U.S. at the lead—have given rise to intelligent criticism of much of the population-control project and sometimes frantic resistance, especially in some of the targeted poor countries.

When one considers together the expansion at 4 percent a year of the population of Kenya and the disappearance of every developed people; when one thinks simultaneously of the twenty-three million people in Mexico City and the quarter of Uganda's sparse population that will die of AIDS; of the annual addition of the equivalent of the population of Canada to the already over one billion citizens of China, while in Toronto three out of four citizens are either first- or second-generation immigrants, then you see that in our reflections on the HTX we shall have to face not just tough demographic issues but newly deepened questions about *Die Stellung des Menschens im Kosmos*—"The Place of Mankind in the Cosmos," to borrow the philosopher Max Scheler's title.[29] That is why in even a most preliminary glance at the HTX like the present one, we shall be confronted by moral dilemmas.

---

29. The Italian population—to pick the worst case—is destined to decline by over one-third in each of the coming generations, if birth rates stay steady.

The Four Worlds

"THE world" is, of course, not just one great undifferentiated glob of humanity. From families to nations, people congregate in social structures of many kinds. We hear much talk of our existing in a very large category at the planetary level: four huge "worlds" characterized by different stages of economic development and manifesting great distinctiveness in their cultures. Here is a good example of the dialectic of the notional and the real: The tribes, villages, and mostly shanty-town cities of the Fourth World of course are objectively existing social entities, visible through their flimsy "infrastructure." The villagers have their own conceptions of their little local worlds. But the "Fourth World" is first and foremost imaginary, existing in First/Second World minds. However, the concept of world on this level is beginning to mirror an objective reality in the Third/Fourth Worlds in the form of societies troubled in their history by outside currents of "development." Third World societies merit their appellation by virtue of already having been dragged into some pretty serious involvement with HTX realities, the fabric of their everyday existence having been fundamentally remolded by the HTX, and they are finding themselves compelled to interact with one another more. The vast networks and "infrastructure" of the First/Second Worlds have passed far beyond the imaginations that continue to direct and drive them into the future. They are tied together into megalopolises by concrete networks of road, rail, pipeline, electric grids, telephone lines, and air service. So while the "Fourth World" as world exists mostly in Western imaginations, the Third World has been defined through connections with the HTX and is now becoming a real world, or series of worlds, the First and Second Worlds enjoy great objective reality. Just how "real" can you get? The HTX molds worlds less the farther they are from the HTX spiritual center in Europe, America, Japan, and the Tigers.

Blocs within the Worlds, and the Two Great Orphans

These "worlds" are obviously not homogeneous wholes. Each world encompasses a number of nation-states, each with its own constituent institution—the government—and these are of various kinds. Other institutional arrangements, and networks between institutions, subdivide each world into sectors like the banking and monetary system, the petroleum system, and the national power grid, each a little world of its own, interconnected with many of the others. At the political level we encounter an important grouping of these national worlds: the *blocs,* constituted by ever more tightly drawn networks of relations between certain countries, sharing common attitudes and conceptions, and often a similar stage of development. Strong commercial relations characterize the blocs. If anything, they are becoming an ever more significant feature of the structure of the world system, cutting to some extent across the various "worlds" and becoming structured by international institutions:

the European Union,

the North Atlantic Bloc,

the North American Free Trade Zone,

the East Asia Bloc (dominated by Japan, which, like the Tigers, is tied in quite intimately with the EU and the North American countries),

the Russian-dominated bloc.

Is there a Central Asian Muslim Bloc? Not really, as I see it, but rather there exists a group of Muslim countries, each formed in its own way by Islam working in the distinctive history of each, with strong variants on the Sunni-Shi'ite tension—stretching from partly Muslim Bosnia through "secular" Turkey (with Islam reawakening), the Muslim former–USSR countries, Shi'ite revolutionary Iran, the Persian Gulf countries with their shaky feudal regimes, the Middle Eastern countries, in tense relations with the highly HTXed Israel, the countries of the Arabian peninsula, medieval Afghanistan, huge and solidly Muslim Pakistan, Bangladesh, and parts of India.[30] Indonesia in the east and the Mahgrebian countries in the west are somewhat off to the side.[31] Saudi Arabia, Kuwait and the Emirates, and Iraq and Iran, because of their oil wealth, play a major role in the world out of proportion to their populations; only Iran has a significantly large people. The tensions tearing through the Islamic world, coupled with the oil domination of the Gulf, creates the SINGLE MOST DANGEROUS INSTABLE SITUATION in the world today. (Only the Taiwan fantasy of China can compete.[32] The disaster in the Balkans is of little danger to world peace compared to this Muslim non-bloc and China, both of which are increasing the danger for everyone by playing sinister games.) The collision of HTX institutions bearing their HTX mindsets with the ancient rich social structures of Islam is proving *explosive*. Perhaps inherently no more disruptive of old being than what is happening in tribal Africa, the HTXing of these countries, because of

30. The largest single Muslim population in any country is over ninety million.

31. That is what the term *Mahgreb* means, and each is quite distinctive with, on top of it, Algeria split by civil war between secularists and Muslim fundamentalists.

32. The anomaly of North Korea is significant only because China has helped that country acquire long-range missiles and nuclear warheads.

their strategic position and their collective oil wealth, makes of this region a prime subject for careful study of what the HTX has there wrought. The weakness of Islam in the area of international organization, the difficulties clear Islamic fundamentalism has in responding to the massive secularizing tendency of the HTX, the recourse to a fanaticized version of Islam by the agitators of the poor masses, the skillful(?) play of the HTX oil-consuming countries, and, the pepper in the pot, the place and role of the highly HTXed Jewish state, struggling to survive in the middle of this pot, is all contributing to the boil and bubble of this witches' brew.

There are two giant countries with no blocs of their own. Both are obviously in their own right massive, difficult elements sui generis in the world system:

India

China

That adds up to five blocs, two nonaligned nation-giants, and one disparate group of Muslim nations. Each is rooted in old religious/philosophic traditions—modernized Western Christianity; glacial Eastern Christianity and a failed Marxism; Islam, Hindu-Buddhist-Confucian traditions—with dangerous tugs-of-war for influence where they overlap, and a particularly perilous struggle for domination in the oil-rich Central Asia and Middle East. Meanwhile, these blocs, the two nation-giants, and the Muslim states are being penetrated and transformed by the planetary being of the HTX, each at its own pace and according to its own character.

Symbol Systems

The just-considered fact that the First World and Second World are symbols for a dense network of interinstitutional play, while the Fourth World is more an abstract concept grouping peoples of low economic development, is a good reminder of just how important—for understanding, for control—and just how tricky is the whole vast human, all-too-human issue of relating symbols, cooked in the rarefied virtual reality atmosphere of the imagination, and reality with which we have to come to terms. The real relationships between interacting institutions and between things making up "productive facilities" and "infrastructure" are themselves not symbols, of course, any more than is my house, real foundation for that dreamier reality, involving spiritual interactions between husband, wife, and children, caught in the symbol "my home."

Each (real) agent of each (real) institution both conceives of, and is told what he has to do next, through symbolic representations and then he communicates his intentions to other agents through linguistic and mathematical symbols, as well as by his actions, "which speak for themselves." The human imagination, reacting to emotions, which fires to action, lives from symbols (and, of course, behind them concepts, the abstract notions that are transmitted through symbols and in terms of which symbols, themselves often dense and concrete in connotation, can be killed by conceptual explanation.) Symbols, from gestures and dress, sacraments, images, to languages, are so much of the stuff of everyday existence, they become so real to us, and are so influential on our relations to nonsymbolic reality, we come to take them for reality itself . . . especially money!

So you see why in *Tradition and Authenticity in Search of Ecumenic Wisdom* "symbol systems" were identified, along with "institutions," "traditions," "processes," and "events," as one of the five most essential elements of "the World System." Our acting together depends not just on the analytic distinctions and the conceptual uniting of many individual things into abstract classes so we can relate them in our minds and deal with them *en masse,* but also on *the emotive efficacy* of symbols: we have to *want* to act together! That e-motional (literally, "moving us out from") capability includes the symbols' invitation to celebrate: the society celebrates itself through them ("America the Beautiful" with the Stars and Stripes unfurled), and the symbols themselves become a means of play and admiration, which we call "art," "liturgy," and "festival."

Traditions incarnate themselves not only in institutional arrangements, based in inculturated habitual roles, but in their own sets of symbols, through which we grasp what we are expected to do, for example, the scriptural, canonical, and theological language of the Church, or the mathematical language of contemporary physics, or the "IBM Speak" of "the Blue Book."[33] To enter into a technical field is in part to "learn its lingo," but it is also to learn the techniques, the "stances," the body language, the peculiar (and always symbolic and meaningful) ways one relates to others on and off the job, which vary from institution to institution. Institutional forms and especially institutional desiderata mold the symbols in which these ends are expressed to the participants. But the resulting conceptions and the poetic dis-

---

33. By "language" here I obviously mean more than a lexicon of terms, and more than the extant literature of the tradition; there are also liturgies, symbolically formed mind-sets, which flesh out ways of being in the world together.

covery of the effective symbols, by forming the overview, return to re-shape the institutional arrangements themselves, along with the reality of the unfolding events and processes of change. A manager, to move his subordinates, must go beyond a bare-bones speaking ability in their language, even beyond knowing "the lingo": He has to know just the right way, in a given culture within a tradition, to express delicate matters to get the best performance from his workers. And he has to know how to conceptualize what has happened and mold new symbols to move the organization forward to deal with unfolding processes and events. (Skilled symbol manipulators are among the most powerful people in society. A common failing in CEOs of the past was their inability to understand and to use symbols effectively. Such manipulative use of symbols is not, however, HTX, just human. In the HTX we have become perhaps more explicit about techniques and more cynical in their manipulative use.)

The various symbols systems move us in quite different ways. Consider several instances:

—The vocabulary of the modern First World state, seeking to express ideals, is drawn predominantly from the constitutional vocabulary of liberal democracies, with Marxist rhetoric on the decline, even in the ill-fated Second World.[34] Legal systems, expressed in legal language, channel our action, as regards commercial law, in formal settings, but like the symbols of various forms of political rhetoric, including today the politically correct New Speak, the symbols of criminal law influence our daily behavior.[35]

—Physical science and technology are based in mathematical forms of analysis, expressed in mathematical symbols, grounded in a way of analyzing the world that goes far beyond the techniques of everyday common sense, which is one reason scientists are said to "live in a world of their own," and only gradually penetrates the popular consciousness, mediated through the poetic symbolizing of journalists and high school teachers. In the biological sciences, symbols borrowed from the Greeks mixed with barbarisms invented by the poetry department of pharmaceutical companies are still prominent.

34. The confrontational, atheist, hedonist mind-sets that drove much of Marxist and Neo-Marxist rhetoric are still at work, of course, in the First World political correctness games of language transformation.
35. The tremendous fight that has broken out in the Catholic Church over the use of inclusive language has grown into one of the most informed and subtle dialogues over symbols in the whole HTX!

—The religious traditions develop theological symbols and liturgical, ritual, and ethical forms that partially fashion the everyday lives of billions of people.

—The symbol creation and manipulation of the gurus of mass entertainment, through a great complex of institutions, is a vital reality built on manipulating imagination and now virtual reality simulation and cyberspace. This is a critical playground of symbols, essential to the war for souls, worthy of a monograph of its own. (Christa Mavis said, correctly, but perhaps slightly exaggeratedly, "We no longer live in a democracy but in a dictatorship of the Mass media (*eine Medien Diktatur*!)"—a phenomenon of symbol manipulation to political and ethical ends of colossal proportions.[36] I mention this war of symbols here because such wars are endemic to the HTX, given the high degree of explicit analysis to which its being has led.

—An effort is being made to develop a set of cultural symbols, driven by secularist humanist ideology, culminating in the self-conscious effort to express a new (or old pagan) catholicism as "New Age." Ranging from Gaia mystique, Wicca, and "Rock" music to Disneyland Mickey Mouseness, massive sporting events, brand name, new, unnourishing but symbol-charged foods (Big Macs and Cokes, symbolic the way champagne and pâté de fois gras have been for the nouveaux riches), all are part of this symbolic effort to uproot and transform masses to various manipulative ends, (if only to improve McDonald's bottom line). Meantime, despite all this HTX playing around, two other ancient catholicisms with powerful symbol systems retain a hold over large segments of mankind, Christianity and Islam. Both are actually expanding vigorously, and while Christianity has lost many in the First World, even there it remains, in the very heartland of the still-secularizing HTX, innovative, forging new symbols and new ways to communicate the ancient symbols. The Muslims seem to me less creative when it comes to inventing fresh forms.

At long last, even the captains of industry, who, busy and pragmatic, have long left symbol manipulation to the professionals—the professors, lawyers, entertainers, politicians, journalists, and public relations and advertising experts—are waking up to the extent to which it can affect the bottom line.

Neil Postman's *Technopoly: The Surrender of Culture to Technology* is, among other things, a brilliantly symbolized indictment of "symbols

36. Graf von Brandstein-Zeppelin, speaking to a meeting of the Association of Catholic Entrepreneurs in Fulda, September 25, 1994.

gone mad": symbols without context, which are, in the HTX, "devouring the psyche."

> Somewhere near the core of Technopoly is a vast industry with license to use all available symbols to further the interests of commerce, by devouring the psyches of consumers. Although estimates vary, a conservative guess is that the average American will have seen close to two million television commercials by age sixty-five. If we add to this the number of radio commercials, newspaper and magazine ads, and billboards, the extent of symbol overload and therefore symbol drain is unprecedented in human history. Of course, not all the images and words have been cannibalized from serious or sacred contexts, and one must admit that as things stand at the moment it is quite unthinkable for the image of Jesus to be used to sell wine . . . On the other hand, his birthday is used as an occasion for commerce to exhaust nearly the entire repertoire of Christian symbology. The constraints are so few that we call this a form of cultural rape, sanctioned by an ideology that gives boundless supremacy to technological progress and is indifferent to the unraveling of tradition . . . Advertising is a symptom of a worldview that sees tradition as an obstacle to its claims.[37]

Symbol manipulation has indeed gone wild in the new HTX. We are confronted with powerful new phenomena, from the virtual reality, science fiction creations of *Star Wars,* and the vile lyrics of the most Satanic rap music, screamed by hideous faces trying hard to look as evil as possible, through to the honeyed tones of the announcer accompanying seductive images in TV car commercials, all requiring careful reflection—not more flights of fancy à la Marshall McLuhan at his most exalted. (The insalubrious game played in the Lewinsky affair by the White House "spin doctors" and the media, utterly leaving the old fogies from the House of Representatives in the dust, achieved a new summit, making the unlamented *Reichsministerium für Propaganda* look like Boy Scouts next to Dorothy in Oz.)

On the Internet we interrelate uniquely through symbols—there is not even the (electronically altered) sound of a human voice. So many of these powerful new HTX symbol seductions draw us away from full, direct, involving human contact, from having and raising children, from contemplating nature directly, even from hands-on dealing with managerial problems, as elaborate business school conceptual dreams get in the way, in the words of the *Economist:* "the 'five forces' of com-

37. Ibid., 170–71.

petition and the 'seven S's' of strategy with fearsome flow charts and complex formulas . . . robbing business of its human drama and ignoring the links between corporate decisions and everyday life."[38]

Institutions

Since the time culture moved beyond a level of animallike *sympathie,* expressed symbols have been the medium necessary for full human cooperation. Although many of our daily interactions are punctual and spontaneous—a little friendly interchange, a small suggestion—many more are molded long-range through our inculturation into roles we play in different institutions. Some of what is expected of us in these roles is explained in symbols. Complex, large-scale social organization would be impossible without writing, and now impossible without LANs and WANs to whiz the symbols around.[39] Institutions have always had to invent ways to whiz instructions and motivations about. How, at what speed, and to whom they get transmitted affects and is affected by the nature of the symbols themselves. Electronic signals and illuminations on parchment are not quite the same thing; neither are an e-mail message and a conference call!

But most of the traditions incarnated in institutions are learned by imitation, observation not mediated by explicit spoken or written symbols, significance being read in gestures-in-situation. Imitation requires *living the truth* of that institution in personal contact or at least through a teleconference. The family, the Church, government agencies, the corporation furnish our "horizons of interpretation," flesh out our vocabulary, and mold our strategies, hence the endless discussions around the dinner table, committee meetings, and the need for apprenticeship in one form or another, in all of which the unwritten behind the written (and spoken) is being picked up.

The institution remains the warp of the HTX. Even a rapid survey of the genera of institutions essential to the functioning of the world system produces striking evidence of their extreme importance in our lives. In some we play roles central to our lives: One is father of this family and district sales manager of this steel company. He is also client of the supermarket, treasurer of the condominium association,

---

38. A good business book that avoids using these as an excuse for "flimsy research, one-sided arguments and banal conclusions" is Rosabeth Moss Kanter, *World Class,* reviewed in the *Economist,* September 9, 1995, 85.

39. Local Area Networks (LANs), link computers in the office; Wide Area Networks link LANs over vast regions.

client at the local bank, and consumer of water from the local company. The new HTX institutions collectively are not just involving but overinvolving us!

As societies are penetrated by the HTX, one can see a certain order in the creation and strengthening of particular kinds of institutions. This order, despite some leapfrogging, remains basically the same for each society moving from Fourth to Third World. Mapping this chronological order of consistently unfolding, typical *essence* of developing societies reveals important aspects of the essence of the HTX itself, as well as being of use for those who desire to plan and manage the development process today—provided they see clearly the nature of the present new explosion of institutional innovation and interconnection. The accelerating interweaving (imagine a vast ball of twine increasing like a snowball as it rolls down a hillside of threads) of an immense variety of institutions with their myriads of distinctive traditions, at once talking both their own language and the linguae francae of the HTX, creates a constantly readjusting whole in which billions of actors must creatively adapt to shifting change, a change that is accelerating and becoming more radical.[40] The Internet, our pet symbol of the new HTX, evokes the kind of openness, the freedom to jump at once to anywhere on this vast ball of twine, suggesting almost infinite pluralism.

But the individual is not infinite, and he has to be formed to handle this HTX challenge, to manage his life so as not to be overwhelmed by it. Many people are failing. The risk that tens of millions within the most HTXed societies will be unable to keep up, and so will have to be sustained by yet another institution, the welfare system, or will die prematurely of stress-related diseases is increasingly apparent. So is the danger of some kind of mass nervous breakdown as we all wear out adjusting! After all, we evolved at a molasses pace until the first explosion, a mere 7,000 years ago: civilization! Just how flexible are we? These problems account for the fact that to the HTX scientific, educational, industrial, commercial, financial, and governmental institutions have been added progressively the mass of private and public welfare

---

40. The actors in these different worlds are called upon to adjust to the demands of other worlds, which come to permeate the horizons of the more englobing traditions. For instance, the railroads have had to adjust to the demands of just-in-time inventory now used by auto manufacturers or risk losing all that business to truckers. This, in turn, has made them dependent on high-speed information processing, so now IBM plays a role in the life of the Union Pacific bigger than the locomotive builders, something no one could have imagined forty years ago, when I was still pounding a telegraph key on the steam-powered Frisco Railway!

institutions, therapy and psychiatric institutions, and "get away" spas and love boats. Huge government welfare programs—which most observers, no matter their ideological tendency, will admit are necessary—are proving "inefficient" in counteracting the dehumanizing pace of demands imposed on us by the new HTX, and impossible to keep financing at present levels.

Traditions

The HTX is woven from the thousands of sectorial traditions that form cultures which mold the work of different kinds of institutions. Each institution both develops its own tradition, and is carried forward by it (dialectic again!), but is also illumined by the being of traditions and cultures broader than its own, affecting the interpretations of actors in, say, other companies within an industrial sector (e.g., oil and gas). The traditions of Royal Dutch/Shell reflect and affect the more general horizons of the industry. Then there are a certain number of widely englobing traditions, carried by vast governmental and educational and information institutions, including the Church, some of the language of which is spoken by all who have been seriously encultured into the HTX. A few are so basic, so illumining of the whole planetary scene, we may term them *foundational-epochal traditions,* passed on by no one all-englobing institution but rather by the weave of vast social governmental and educational institutions. They have become aspects of the being of the epoch. Within the horizons of these very broad worlds the many *cosmiota* take their place, and receive some of their sense.

With characteristic modesty, the HTXRG brings you this catalog of the foundational-epochal traditions at the base of the HTX:

1. *Bourgeois-liberal.* I do not like the Marxist contamination of this symbol. Yet the complex weave of developments growing out of the rise of commerce in Europe, contributing to the growth of cities (*le bourg*) and of markets and banks, demanding the development of mind-sets, including a certain notion of *liberté* and "the rights of man," and eventually demanding the "revolutionary" establishment of a structure of new laws to guarantee the stability and safety of commerce and the "pursuit of happiness," are probably still best labeled "bourgeois-liberal."

The liberal outlook is invited by the experience of being responsible for one's own accomplishments in an open setting lived out by the daring, far-traveling merchant. He is dependent upon his own initiative and hard work, not upon his birth into a feudal status. (You are what you make out of yourself—what you do—not what nature [or Providence] has given by birth. So we meet again *doing versus being, will-to-power versus gratitude and reverence*). To secure space for himself, the bourgeois had to admit the freedom of movement for others, thus making it impossible to control thought centrally and eventually leading to the recognition of freedom of conscience and the acknowledgment of certain fundamental rights possessed by all.

The bourgeois-liberal tradition is becoming increasingly multicultural. That means, within liberal society, an inevitable relativism of what are called values.[41]

What the fundamental rights will be is determined by the common conscience—hence the rule of democratic (as opposed to feudal) law becomes essential, "*la société civile,*" secularized, independent of divine-ecclesial authority, in the extreme, "English common law," pragmatic and relativistic vs. *sha'aria*. I put it this way to point out the dilemma we have seen facing Muslim countries: Will Islamdom be able to survive the Enlightenment, the transition from divine to liberal law and still remain Islam? Christendom did not, but, arguably, a Christianity true to its deeper self can continue to survive the disruption brought on by the wrenching transition from feudalism. Whether democratic rule of law as a genuine protection of the individual from abuses of manipulated state power can long survive is a burning question in the HTX. Was the fall of Germany into Nazism a passing anomaly, and of Russia into Communism a failure to make the transition out of feudalism? More on that later. An enlightened Western view today could reasonably hope as follows: Building on a residual sense of the Christian revelation of God's love for every person, the need to protect rights should lead to the sense of the intrinsic natural worth of every person, and then to reflections on the need to make freedom effective by providing adequate opportunity and an even playing field.

One of many signs that this hope may be infected with mere optimism: The fact that "rights language" is now being abused by groups not in the least interested in the common good (which still motivated the Jeffersons and Madisons of the less aggressively atheistic branch

---

41. The notion that values are projections of what one wills distracts from the earlier notion of objective goods that are inherently what they are whether we recognize them or not, or whether we have the good sense to pursue them or not. To show how far this kind of arbitrary sense of values ("You have your values, and I have mine") has penetrated, when papal encyclicals are translated into English, *bonum* is rendered "value" by translators who obviously do not know any philosophy.

of the Enlightenment) poses serious dilemmas for liberalism.[42] A wide range of philosophies can be grouped under the "liberalism" banner, ranging from a sense of the need for individual responsibility, all the way to an adolescent libertinism, which can be very close to the illiberal voluntarism of *dirigiste* ideologies like Leninism and Nazism. The claim here is simply that liberalism in some form has been, and continues to be, an essential ingredient in the development of the HTX. That the sense of this liberalism is constantly evolving and that it is everywhere under pressure seems obvious.

2. *Progressism.* Secularization of the Judeo-Christian eschatology—the revelation that history is unfolding toward a fulfillment, willed by God—is entangled with every form of liberalism. (What is the point of taking big risks and working like a dog unless you are convinced you will improve your lot?) The HTX is driven by faith that generates *savings,* and hence investment. Sometimes this belief will go as far as faith in the perfectibility of man, even to the extreme of Nietzsche's "Man, save yourself!" and dreams of the "Overman." ("Designer genes" are on the way.) But often it is no more than the diffuse optimistic sense (better supported by the evidence than "perfectibility") that human ingenuity, coupled with some good will, can achieve the *material* betterment of mankind.

3. *Scientific-technological traditions.* The technological mind-set is much broader than just attitudes acquired from the scientific traditions; it involves the growth of discoveries and habits of thinking in management and government. Certain basic aspects of technological mind-sets have become diffused throughout the horizons of the HTX.

4. Interplay between the three above kinds of traditions has fed a *tendency to immanentism, to the sense that "everything can be changed" ("Pelagian"),* feeding a practical atheism because it tends to erode the sense of beholdedness and gratitude for what a providential God has intended through the sacramental gift of His creation.[43]

5. *Traditions of entrepreneurial-social welfare political economies.* Spawned

---

42. I always say that the supernatural virtue of hope, founded in a sound faith in a real God, is separated from mindless optimism by a thin line!

43. This tendency to atheism is resisted, of course, by traditions nurturing a responsible liberalism, especially by religious traditions fostering the sense of dependence on grace, including the gift of the cosmos with its natural human forms. If liberalism equals voluntarism, then a Christian liberalism would entail a heterodox, ungrateful Christianity. But if liberalism equals a sensitivity to maturing the freedom of the individual, to allow him to become an adult responsible for himself and with a rightly developed conscience, then indeed a Christian liberalism would be possible and healthy for the society.

both by the liberal mentality and certain HTX managerial attitudes and conceptual abilities, these traditions, mixing entrepreneurial openness with the desire to provide a social safety net, grow through phases. The democratic entrepreneurial-social welfare state known today in the First World appeared in its present form fairly late in the development of the HTX. But still I believe the mixture of market-openness and safety net has become an essential feature in the HTX: it affects the powerful, wealthy core democracies and, through them, the world. It proves a pole of attraction for the development of other societies.

Not a single First World country of Christian origin is without this mixture of attitudes, institutions, and laws encouraging entrepreneurship and an expensive bureaucratic social welfare net. Japan lags, both in openness (from within and without) and in development of a net, although its industrial policy, which up to now has virtually guaranteed full employment, has made it less necessary. Some of the Tigers (especially Singapore and South Korea) have been quite *dirigiste* and protectionist, and none of them has as yet built a very elaborate safety net. Will they? Or will they depend more on intact families and large saving rates, and thus economize on the huge costs of inefficient social welfare bureaucracies, which will give them a continuing competitive advantage? Or will the HTX erode family cohesion there as it has in Europe and America, so that wealth will lead to consumerism and seriously declining savings rates, both of which will increase demand for state aid to pensioners? As the emerging Third World countries and giants like India and China become increasingly dominant competitors, their economizing on safety nets creates pressure to scale back and privatize social welfare in America, and maybe in old inflexible Europe, as *Festung Europa* may prove too costly.

6. *Socialism, or central command–egalitarian ideology.* The collapse of the Soviet Empire and the inroads of a raw entrepreneurial capitalism in China by no means spell the end of the tension: socialist state–assured security versus free-market competition. The dynamic liberal market economy is unbearably hard on large segments of the population. We witness today the slide of Russia back into *dirigisme*. The temptation to "humanize" the system (i.e., impose some attention to basic needs) will remain, despite all the evidence of the impoverishment it brings in the short and long run.

All-Englobing Notions in the HTX

Running through the horizons of interpretation of these vast diffuse traditions are even more basic and omnipresent notions. They are quite recent concepts, and it is astonishing how such new notions could have become so basic so fast. They offer glimpses of certain fundamental processes at work in the HTX and hints of widely inculturated mind-sets. Through these we get a first glimpse of the *essence* of the HTX.[44] Symbols of some of the notions are commonly recognized; others have become *too* common, so taken for granted the proverbial fish fail to see the liquid ambience. But the worldwide society nevertheless swims in it and drinks it in!

1. Progress and Planning. Take for instance the very notion of "progress." Does it not entail a particular notion of *planning*? The squirrel plans for winter; the tribesman plans his hunt. But the primitive agriculturalist, because of his technology, will be obliged to lift his sights, to plan further ahead, to save seed for the next season, and to observe the changes in seasons and weather on a different scale and with greater finesse than the hunter. But with the notion of a *pursuit of progress,* and long-range planning, come those questions we have seen being raised of "to what end"? "by what criterion of the 'better'"? And as goals become more explicit, longer range, and in other ways more ambitious, "resources" are seen to be needed, requiring *management* on a different, more complex scale. As enterprises expand to faraway continents, and as notions of long-term investment, market share, and return on investment grow, so do notions of design, production, raw materials and parts procurement, after-sale maintenance, marketing, and advertising as a part of marketing, all aspects of HTX "management." Bubbling clusters of worlds within worlds and overlapping worlds on the same level and scale of activity frothing up effervescently is a manifestation of radically new mind-sets and of new intensity in the development of certain older mind-sets. "Progress" becomes a fanatic pursuit of "self-development," "growing the company," "economic development," "growing the economy," "reengineering," and other symbol-garbage, all revealing a rapid shift in being. From the real everyday, strategic problems of managing bigger, more far-flung enterprises arise reflections of centralizing versus decentralizing, of hierarchical control versus assigning responsibility closer to the action, ("de-layering"), of reducing overhead and downsizing—all relatively new HTX concepts, each flying its sexy verbal symbol.

2. Efficiency. Underlying all of these notions, the *ontological* concept of *efficiency* emerges, and a *calculating* efficiency mind-set permeates

---

44. Yes, an epoch can have an essence, but this is not to be confused with the question of whether the HTX itself *as world system* may not yet have achieved (or may never achieve) a coordinating social organization, so that the emerging form may never come to something like closure, the world system will remain very ragged and the epochal essence fluid.

the society: the idea that it is desirable to achieve a given objective with the least wasteful expenditure of resources, especially time—"Time is money!"—of intelligently ordering resources to achieve what you want economically.[45] Hidden in the concept are fundamental presuppositions about the priority of *doing* over *being,* that somehow the "really real" lies ahead, in the yet-to-be-realized project, which has to be reached through the expenditure of the least from a past accumulation of finite possibilities (like "this ore body" or "this aging work force"). To see how tendentious is the ontology assumed by this founding notion, compare a nature trail to a highway: the nature trail is never the shortest distance between two points, because it exists to be enjoyed, it wanders, inviting the hiker to linger in delight at the sights. The highway is bulldozed through mountains and even through town centers. (And yet I know HTX types who have only one idea in mind, "to *do* the whole Bruce Trail, all 760 kilometers of it," as though the object were to get somewhere instead of enjoying already being there!) When the vertical transcendent through which the eternal reaches down to us is ignored, the eschatological future that has been revealed as a fulfillment of "the Kingdom" already among us gets deformed into a never-ending progress that cannot achieve a proper present. Even the spiritually minded become fanatic doers of "spiritual exercises" that are supposed to produce efficiently faith experiences.[46] (More on that in chapter 6.)

3. Information. Since Adam and Eve decided to become gods by eating of the fruit of "the tree of the knowledge of good and evil," any human being pursuing any goal has had to gather adequate information.[47] (Our First Parents' information about future possibilities turned out to have come from an untrustworthy source!) A million years later the concept *information* at last becomes explicit, and "information" (united, oddly, with entertainment) becomes the largest industry.[48] From the beginning of "business," bookkeepers, and later accountants were, in fact, information gatherers and processors, though no one thought of them in those terms. Now "IT" departments run by CIOs

---

45. This is a notion utterly foreign to all previous epochs in all traditions; think of the medieval sculptor finishing to perfection the back of a sculpture that will stand high on a cathedral facade.

46. To what extent were St. Ignatius of Loyola and his Jesuits, who transformed modern education in the seventeenth century, already touched with elements of an HTX mind-set?

47. Their problem, in addition to an uncritical, proud, unrealizable goal, was manipulative disinformation, provided by the enemy, about the competitor and the tree!

48. The start of conceptualization can be dated precisely: Claude Shannon's celebrated papers during the Second World War on communications theory.

(chief information officers) in huge firms can themselves be unmanageable shops. Current wisdom says that gathering and analyzing good information and getting it to decision makers in forms they can use is *the* key to the survival of any enterprise.

Neil Postman, in *Technopoly* believes this is leading to a disaster. He defines Technopoly in terms of its "information immune system being inoperable," adding

> Technopoly is a form of cultural AIDS . . . Anti-Information Deficiency Syndrome. This is why it is possible to say almost anything without contradiction provided you begin your utterance with the words "A study has shown . . ." . . . It is why in a Technopoly there can be no transcendent sense of purpose or meaning, no cultural coherence. Information is dangerous when it has no place to go, when there is no theory to which it applies, no pattern in which it fits, when there is no higher purpose that it serves.[49]

The sense that information is vital has grown explicit because the being of the HTX encourages refined calculation of means to achieve goals, requiring better information about resources and processes (trends), and more explicit reflection about management structures the better to control relevant processes and to anticipate, or at least to react rapidly to *events*. It is this preoccupation with information that gets out of control, inviting failure to see in any depth what is being revealed in the flood, missing the ultimate—and hence the real—context, especially the true depths of the *present*, because one is obsessively absorbed in the pressing and immediate (the surface of the narrow "present"—"the urgent" hiding the Real Presence) being generated by the extreme openness the informational revolution has caused, precisely "Postman's Complaint."

The gravity of this disorientation through our being overwhelmed with ill-digested data is gaining recognition. A recent Japanese management best-seller promotes the difference between companies that are able to absorb data into strong traditions of interpretation, thus converting it into usable information, and those that are submerged by it (and waste a lot of money gathering it!).[50] We shall return to this central HTX dilemma, as it affects all our lives.

49. Ibid., 63.
50. Hitosubashi University scholar Hirotaki Takeuchi, *The Knowledge-Creating Company: How Japanese Companies Create the Dynamics of Innovation* (New York: Oxford University Press, 1995).

So now we have glimpsed something in all the elements of the world system recounted by *Tradition and Authenticity* (chapter 6):

—*institutions* develop to achieve leverage over

—*processes* and to anticipate and react to

—*events,* using

—*symbol systems* in analyzing, storing, and communicating needed information,

—in the service of *tradition* (continuity in the pursuit of a vision), and further formative of it.

All these elements interact, in a never-ending evolution of any world, small or all-englobing. Challenges confront us in the HTX with every one of those elements, and in all their characteristic interactions. *Processes* have become more complex, more dynamic, and gigantic, indeed often planetary in scale; the *institutions* needed to control them are often insufficiently aware of what is essentially involved in many processes and run along too far behind *events;* they become problematic bureaucracies, the top ever further removed from the real world where action has to be taken; in desperation they are perpetually "rethinking" and "reengineering" themselves. *Tradition* remains formative but gets obscured by the uncontrolled flood of information about events, harder to fit into any pattern of understanding. This invites obdurate clinging to old ways of doing things (bad traditions) rather than sensible renewal of traditions, drawing on their profound resources. *Symbol systems* multiply and become enriched in exuberant and confusing fashion, without all this leading to more profound reflection, au contraire, the resulting noise and confusion makes finding time and inner peace for reflection increasingly difficult.

"Being" versus "Having": "Instrumentalization of Man in the HTX"

The Revolt against Instrumentalization

In the explicit goal- and efficiency-driven atmosphere of the HTX everything becomes an instrument. Nature, to be sure, is there, but now viewed as resources to be exploited, even the people involved become "personnel" or "human resources."

Not everything and everyone succumbs graciously to this abuse. Worlds, with different being, collide, as we see, for instance, when environmentalists revolt at nature being considered just a gratuitously available storehouse (and dumping ground for waste). The Church incessantly preaches that persons are not to be treated as means.[51] And worlds collide when a decent human being cast in the role of manager worries about what effect moving workers about will have on their families.[52] Some tough managers may be reluctant to admit that they are actually being a bit humane; we are not generally skillful at reasoning between, and balancing the goals of, worlds with different ends. Competing eschatologies requiring different notions of "the Kingdom" are at war within the same person. "Man cannot serve both God and Mammon," we were long ago warned.

As both information and management technique improve, the manager can reflect, "I have enough to do juggling all these balls without attempting to play God with my employees and customers. Accounts receivable from a certain customer known to be in trouble have become too large. If I stop further delivery, this may be a death blow to his little company. It is bad for me to lose a customer, and if he folds, I will not be paid back. But should I accept further risk for my company just to help his company, and by keeping it afloat, help our community?" Tough managers and tough banks have been known to be human at times, to assume additional exposure when a narrow business perspective would indicate cutting their losses. This will most often involve friendship, for two reasons. The first is practical: if the banker knows the manager well it is easier to judge the quality of the other company's direction, hence the manager's ability to see the troubled company through. The second is human: the loyalty one feels for a friend. Loyalty may sometimes contribute to a bad business decision. Conversely bad business decisions that destroy employee, customer, or supplier loyalty can be terminal. This is one of those large contexts where we see that the HTX is not built on unfeeling machines but on a base of quirky, "fallen" human nature. "Personnel," "suppliers," "customers," "government," and "competitors" all have *personal,* very human agendas that go beyond the efficiency demands of the political economy. Even the dentist cannot reduce me successfully to a mouth full of teeth—a scared, irrational

51. Scott Adams, the author of the Dilbert cartoon, once again makes an HTX mind-set look ridiculous: The pointy-haired boss assures Alice that he will add "resources to her project." Ungratefully she screams, "WE'RE HUMAN BEINGS, NOT RESOURCES!" To which the boss replies, "Would it help if I told you that resources are our most valuable asset?"

52. IBM Canada has just created a new managerial position at the corporate level; the person in it is charged with "environment, health, and balancing family and work requirements," which they call (UGH! Call Scott Adams) "well-being."

human history comes attached to those molars, and if he does not watch out, I lose confidence in him as a human being, excellent technician though he be. Will this guy go out of his way for me in an emergency?

A vast amount of business is conducted on *trust;* and trust and loyalty are difficult to separate, indeed they feed on one another. They both have something to do with human goodness, which is not basically driven by efficiency. One could put it a bit cynically and acknowledge that the HTX manager who is *really* efficient in the long run, and not just obsessed with the immediate bottom line, will integrate human nature into his equations; he will seek not just narrowly competent people but good people with whom to work, building relationships of trust and confidence, which cannot happen in the absence of all loyalty.[53]

In a recent book about the structure of the new "virtual corporation," William H. Davidow and Michael S. Malone, two certifiable non-cynics, declare themselves convinced "trust is a defining feature of a virtual corporation."[54] One of the characteristics of the corporation woven together with customer and supplier by computer is the high degree of participation of all—designers, part suppliers, and customers—in the conceiving and production of the product or service. This requires sharing proprietary information that before would have been carefully guarded. Confidence that all contributors will protect the information is a sine qua non.

Time and Presence

The HTX is often accused of being inherently dehumanizing. Consider *Langan's Law:* "The quality of being *efficace* is inversely proportioned to the quality of being *sympathique.*" It is partly a question of time. The *sympathique* fellow is absorbed in the moment, fully present; there is no tomorrow, and so he is always late for the few appointments he is obliged to make. The efficient Yuppie is always looking over your shoulder at a cocktail party to see if there may not be someone more useful in the room, because he is *using* every minute (and everybody) to advance his cause, whatever it may be just then.

"Time is money." As soon as a full sense of "investment" becomes explicit, and one realizes that "return" is measured in discrete time segments, then "every minute has to count."

---

53. The CEO of one of America's largest supermarket chains—sales $24 billion—told me he spends half his time counseling his senior officers about personal and family matters for which they've asked advice.

54. William H. Davidow and Michael S. Malone, *The Virtual Corporation* (New York: Harper, 1992), 9.

That is indeed dehumanizing. But why? Well, what does it mean to be human? We have mentioned the existentialists' objection that the HTX ignores the distinction between being and doing, but Gabriel Marcel also distinguishes being and having, *l'être et l'avoir: being present* has to do with fullness of gift of self, intensity, honesty ("full disclosure"), loyalty, trust, generosity, and consideration, which combine in one symbol: love.[55] Of course, one has to "*have* time" or "*take* time" to be with a friend, but "spending the time" is a necessary and *not a sufficient* condition for *being with* the other. And when you really are there, you do not count the minutes.

With the emergence of the postindustrial, third-generation IT-driven corporation, the whole sense of time and of presence is changing. Is the New HTX beginning to reverse the dynamics of the Old HTX, turning back in the direction of *sympathie*? Some have claimed nimble small companies are running circles around the big ones. Cheaper communications make it possible for them to be present in the wider world, but above all, it may be because, through flattened hierarchies and "work teams," coupled with humanizing innovations like flexible hours and working from home, they are more *present* to their associates and customers, more conscious of their needs.[56] If they do in fact really get closer to them, is it possible they are being forced to acknowledge a fuller spectrum of needs, those of the real person and not just the narrow needs of a particular abstracted function?

The spiritual reality of various intensities of attention, of appreciation, and of compassion manifests a raising up of brute bodily energy to a new kind of being, which often calls for firing up brute energy to levels of heroic endurance. There is no quality difference in "electrical energy" derived from solar panels or from fission of U-238 producing heat to boil water to produce steam to turn a generator, whereas in *human energy* quality is everything, once a basic needed quantity is metabolized and transformed by brain and spirit into a transcending form.[57] It is impossible to calculate quality and quantity of "psychic en-

55. This is the expression chosen by the Catholic existentialist Gabriel Marcel as the title for one of his greatest books.

56. That is the thesis of Davidow and Malone. The matter is not so simple. Well-managed big companies, with skillful use of IT and modern communications, with vast networks of influence and deep pockets, can still win out over many smaller competitors.

57. I explored the uniqueness of psychic energy in *Being and Truth*, chaps. 6 and 7. I there confess to misgivings about this notion, but I found it difficult to avoid for reasons explained there.

ergy" and therefore to include it in the analytic plans of an HTX enterprise, and yet, in the final analysis, the success or failure of any human undertaking depends on the quality of the psychic energy mobilized.

The Dialectic of Software and Hardware

Instrumentalizing man may be dangerous, but much psychic energy has always been deployed through instruments—not just instruments of expression (symbols)—but tools of every kind. New notions of what might be accomplished have always enticed man to develop better tools, while the development of the tools in turn has opened new possibilities for seeing, hearing, probing, and sometimes new ways of doing things.

But with the new HTX, has there not been a quantum leap, perhaps an essential change in the nature of the tools, altering the man-tool relationship itself? Is there something in the science fiction notion of tools taking over from man, even making of man their tool?

Stay sober! To be sure, at the moment we are living through an explosion of "information highway" capability that is outpacing our collective ability to imagine very well how to tune a broad band of possibilities into humanly useful wavelengths. Not only is new hardware demanding extremely difficult software development, but the dialectic of both is stretching our collective imagination more rapidly than ever before as we seek to domesticate and integrate all these new capabilities.

Take this example of hardware-software dialectic: "Internet shopping" demands development of some form of "cyber-money" that will be fraud protected. Recently the first "cyber-bank" was established, "Virtual Holdings Corporation" (I did not make that up, it is really its name!).[58]

This is just one minor example of rampant development of one area of software-hardware interaction. Now consider how many identifiable areas of hardware there are, each requiring software. Recall seven genera identified by the HTXRG:

---

58. Basically it is a confidential file of account numbers, which makes it possible for a purchaser of an item described on the Internet to allow the seller to inform the "bank" that it may debit the purchaser's credit card account, once the bank receives the requested confirmation from the purchaser—all this without the credit card information having to appear on the Internet, where some hacker might put it to bad use.

Particle controllers (manipulating various bands of the electro-
   magnetic spectrum of radiations), including
—electrical generators, transformers, transmission lines, storage devices[59]
—x-ray machines, radio frequency transmitters, nuclear medicine machines, microwave devices
—devices harnessing these capabilities for communications:
   telephones, radio, television
—computers

Optical Devices (operating in the visible part of the spectrum):
—eyeglasses (important for literacy!)
—micro- and telescopes
—fiber optics, for seeing inside vessels in the body, and for information transmission, using light waves instead of
   electron beams

Ultrasound Devices
—for cleaning inaccessible crevices in tools
—for probing the body, and for destroying kidney stones

Molecular and Cellular Manipulators
—antiseptic cleaners, vaccines, antibiotics, genetic engineering
   devices
—pesticides, fertilizers, improved seeds

Other Power Sources:
—propulsion devices (steam engine, internal combustion engine, jet and rocket propulsion)

Precision Automated Tools and Robots

Newly Exploited Natural Materials and Artificial Materials
—lightweight, stronger, and rustproof metals
—plastics
—new textiles
—synthetic rubber
—new kinds of molecular structures

Except for eyeglasses and telescopes, every species of apparatus
mentioned has been invented in the last hundred years or so, hundreds

---

59. Do you know how many lines of computer code it takes to simulate (for
training purposes) the control room of a nuclear generating station? Well over two
million. Without simulators—HTX virtual reality–generating machines—it would
be difficult to train operators sufficiently to deal with every possible emergency
that can arise from these, the most powerful concentrations of steadily deployed
energy man has achieved.

of thousands of subspecies in the last twenty-five, some of them displaying quantum leaps in capability!

Each species of new hardware within these genera spawns "software," first as techniques that have to be developed for their manufacture, then for their application; finally, through training, human agents are inculturated into distinctive roles necessary to put the instrument to good use harmoniously within an institution. Typically, software development runs along breathlessly behind hardware development. For instance, Stephenson's steam locomotives were capable of performances for which no adequate dispatching system existed; signaling had to be worked out in a hurry under the pressure of operations. Part of the preservation and transmission of such software is the development of technical symbol systems.[60] A whole new language of train dispatching grew up.

So a dialectic occurs, not only between software to simultaneously exploit and explore existing hardware capabilities, leading to further refinement of them, but as well a machine-to-machine dialectic (mediated through human brains, for now!): The new refined machine capabilities suggest attendant machines to extend and further develop the basic capability.

And so the whole system of machines grows dialectically. The almost microscopic electric motors in a mini-robot are vital, but we almost forget about the (negligible) amount of electricity our mini-robot will use, and think hardly at all of the vast power generating and transmission and transforming system, without which the robot would be deaf, dumb, and paralyzed. And who thinks of nature behind and below the power plant, the source of all the energy being transformed, whose laws have to be obeyed in transforming and shaping all the materials that make up boilers, generators, voltage regulators, transmission lines, and so on? Improvements at any stage along the line may hold potential for refinements at some other stages. To the service of each of these, men have had to be trained and their skill employed in a carefully regulated, institutionalized fashion, requiring information processing, of the resulting symbols.

In the accelerating synergism generated by this elaborate dialectic, man is getting swept up into the breathless new life forms of the HTX phase two!

---

60. Every part of every machine, its every operation, and every aspect of its maintenance has to be caught and manipulated in a symbol, primarily linguistic, but with computerization, also digital. "Software" is symbolic.

Chapter 4

▼

The Heart of HTX Capitalism
Caput, Capital Flow, and Information

What Rules the HTX: Caput

In our quest for the sense of the more magic symbols hinting at the HTX essence, an important truth might be obscured: like all historical phenomena, the HTX—planet embracing though it may be—is still mediated by the fragile individual human soul.

As being illumines individual minds and through them human work gets directed, as being begets itself in humans' concepts and symbols, all this happens through inherited concepts and symbols and in interaction with various realities encountered by the soul. These range from the most objective to the most imaginary and "virtual," all of which present themselves, providing the "ob" (Latin, "over against") in the ob-jects. These concepts, images, and symbols serve not only for appreciating the world in its "otherness," but also as guides for organizing work, so that things get physically transformed as we mold and build them according to our concepts and images. Then in turn, in the form of cultural objects, from tools to infrastructure and as a fund of acquired symbols, these physical and spiritual products of mind help further guide the unfolding deployment of that being, through human souls.

Being in-spires human beings in their use of the traditions—the already achieved languages, institutional arrangements, mind-sets and habits, and tools—to lead orderly, indeed *traditional* lives in society. (Traditions do not have to be old to be effective. Consider how the new cultures of the sixties quickly became massive means handed on for forming the X generation. The dynamic essence of the epoch, founded in "the already having been," assures a certain consistency and order as "history" unfolds.)[1] All this happens in the soul, especially *the vision* and the kind of *character* individual persons are developing, and is central to the being of the epoch.[2] Even what we know of the transcendent-divine,

---

1. Recall *Wesen* = *Gewesen,* equivalent of Aristotle's "what is as already having been."

2. Character is the weave of virtues and vices a person has built up. Aristotle's term got translated into Latin as "virtue" (*vir* = man, suggesting strength), mean-

just as all we know of nature, comes to us through the limits and powers of the human psyche, deposited in the teaching and practice of traditions.

One could draw a mistaken traditionalist conclusion from this, that new being really emerges from "already having been" deposited concrete possibilities. For example, I do not invent the English language as I write; rather I draw on my past learning of the language and its uses, which allows me to write English. But as Heidegger has explained, new being comes to be (obviously, if you think about it) *out of the future,* from the "not yet," or the "no-thing," otherwise it would not be really new at all. The new takes the form of an horizon-opening illumination (in the human soul) of what has already been. Every new breakthrough, even the most modest, throws some new light on what we already knew. But the "already having been" does make a *necessary* contribution, for not just anything can come out of anything; nothing ever comes "totally out of the blue." Our great cultural constructions are all standing on the shoulders of giants. After the fact of a new breakthrough, we can always see a certain continuity between what has just happened—however surprising—and the ground prepared for it, often over centuries, in the "already having been."[3] That "already having been" is rarely just "raw" nature, almost always it is culturally transformed nature, working through previously forged concepts and symbols, and guided by institutional structures and tools without which the new would have been inconceivable.[4]

That is why all of us in the West think and act "HTX." The essence of this social reality has deeply affected all our social structures, and it accounts for the kinds of cultural objects we have to use, from hammers to city layouts and telephone networks, with all of which our souls have to be educated to deal. Nothing appears exactly the same to a modern man since this historical development got under way; even a forest seems to us today either a "timber farm" waiting to be harvested, or a "nature preserve," or, more rarely, "primeval forest full of tourists." And we all act in the ways HTX man has come to act.

Central to this new order is the realization of the importance of that characteristic psychic dimension we have termed *caput,* for the most im-

---

ing a good habit built up in an area of capability (St. Thomas calls such an area a "power"), and vice is a bad habit, the result of a poor habitual use of the power.

3. A genuine miracle is different, of course, for that is an utterly new creation, which, therefore, always remains totally mysterious.

4. The new being is founded on the already having been, but not totally, or there would be no *new* element, hence no future, and nothing really happening!

portant store of capital in capitalist economy is in the head. That is why we invest so heavily in formal education. Whereas for Christian man, the soul was centered more in the heart, so that friendship, loyalty, trust, and, above all, God's love flowing through our hearts were considered golden, for modern man, it shifts to the head. Cleverness, information, and technique are what "pay,"—forget "treasures in heaven." "Pay off" is not grace but prestige and, above all, a high bank balance. Scientific-technological and eventually industrial progress started, not at all in the agriculturally and resource rich lands, but in centers where people responded to difficult situations with *ingenuity:* in the little city-states of Italy, the Dutch and Hansa cities, with difficult "hinterlands."[5]

Still a certain quality of heart—entrepreneurial daring—is valued in the HTX, but perhaps those peculiar qualities of soul that achieve successful bureaucratic maneuvering are as large a factor on the scene, although everyone seems reluctant to vaunt them. The fathers and doctors of the Church valued intellect, too, but, like the pagan, Aristotle, they recognized that the intellect finds solid, lasting truth only when the heart—and character, built on a good heart—is sound. The HTX man is not searching for lasting truth but for ways to exploit the present situation to fulfill his desires. Michael Novak points out that capitalism requires a certain ethical-cultural base.[6] He does not however underline the importance of bureaucratic game playing skills as an essential part of the HTX. The role of manipulation is underplayed because the residue of Christianity in us detects the taint of evil in it.

5. Toynbee showed that great innovation comes when nature is neither too demanding nor too giving. Large populations and poor agricultural conditions forced mass emigration from Europe to America. The immigrants brought habits of determination, discipline, and courage that constituted a "capital," perhaps more of heart than of caput, the fruit of long struggle, coupled with tendencies to think for themselves, attributed by Max Weber to a Protestant work ethic (a *sectarian* mentality, the threatened minority using its wits to fight back). Spain and Portugal failed to develop an entrepreneurial high-tech industry, not because of any lack of hardy peasants or daring explorers, but because their leaders had grown accustomed to living it up on gold and silver from the colonies. Capital for them was stored up in as many palaces as they could command, all intended to symbolize the *grandeur* of those of rank. The colonial elites of the Americas developed as semifeudal exploiters of abundant slave labor and failed to develop early on the *spirit of analysis* that develops that sense of efficiency central to the HTX mentality. Industrialization was brought only at the end of the nineteenth century to Argentina, Chile, and Brazil by immigrants to Buenos Aires, Santiago, Sao Paulo, and Bela Horizonte from industrializing Europe. The colonial transferers of intellectual capital to countries like Egypt, Algeria, and Iraq were never interested in, or invited to, integrate with the natives, who were managed rather than conquered and transformed.

6. Novak, *Democratic Capitalism*, 333–60.

Contrast these attitudes to the Confucian quality of family loyalty that plays such an important role in Asian "capitalism." The Confucianisation of the HTX in Asia is already causing significant problems. The totally instrumentalized HTX enterprise, in contrast, is decreasingly loyal even to its longtime employees; relations become increasingly depersonalized and manipulative, as the struggle to keep the enterprise afloat in a murderously competitive environment leads to increasingly brutal "reengineering," accompanied by intensified power struggles at all levels. That education which is the key to developing HTX *caput* is not just "formal education" in expensive schools, where indeed many traditional symbols and know-how of analysis gets passed on, but, at the beginning of capitalist take-off especially, even more *on the job education,* where people develop skills and ideas and acquire an "it can be done" attitude.[7]

As industrialization moved into sophisticated technology, the formal education of scientists, engineers, and even MBAs (!) became essential—the traditions had to become ever more explicit. But again, it is not just a matter of investing immense sums in postsecondary schools and technological institutes. Probably no countries spend a larger portion of national income on formal education than Russia and Canada. Russia ranks poorly in the competitiveness that is fueled by inventiveness, and Canada lags in patents and Nobel prizes.[8] India produces five times as many engineers and scientists as Taiwan, but their inventiveness has until recently been frustrated by a central-command, bureaucratic government casting gloom over the whole economy. More on education later.

The Lifeblood of the HTX: INVESTMENT

When this intellectual capital develops to the point it enables agriculture and commerce to throw off a large enough surplus, industrialization becomes possible. Part of the intellectual inheritance of a well-developed industrial capitalism is some degree of societal under-

7. The "we have always done it this way" attitude coupled with the employees' conviction "the company is out to get us" of the American railroads where I worked in the 1940s formed a devastating cultural heritage. The private truckers' attitude that "the sky's the limit," their entrepreneurship and hard work, and their ability to exploit the advantages of a more flexible technology devastated the railroads, which only in the last decade have been learning to be more flexible and customer oriented.

8. To be sure, Canada suffers not only from being a branch plant economy, but also because it has developed a baggage of socialist protectionism (though it is not quite on a level with Russia). Most Canadians treasure their social programs.

standing that the fiscal accumulation should not be unduly immobilized in underproductive assets (e.g., purchasing vast forests to hunt) or altogether nonproductive assets (palaces that produce only prestige). Wasting it through mass conspicuous consumption and ill-starred wars of conquest has proven likewise unhealthy to capitalist society.

For capital, in the narrower sense of the monetary abstraction of the fruits of *caput,* to become productive, a significant segment of a society must somehow acquire a modern vision and ingenuity, a whole modern sector has to develop, with all the interacting mind-sets, a "grammar of enterprise," and, for whatever reasons, many producers have to forego too much unproductive personal consumption.[9] But whatever else occurs, the cardiovascular system of the body economic had better operate: The blood of money had better get to the *caput!* When a small elite controls access to capital, as it long has in much of Latin America, vigorous new elements with good ideas starve for oxygen, brain cells die.[10]

Markets can be efficient allocators of capital only if accompanied by good information and the good judgment needed both to know what data to seek in the first place and to act on it. The billions misallocated into ill-thought-through schemes and even outright scams shows that hundreds of thousands of investors are either dreaming or uncritical, unaware of how poor their information is and how skewered the judgment of "security analysts"; they even fall victim to disinformation in a greedy rush to get rich quick. The recent collapse of vast new Russian pyramid schemes, like the now notorious MMM, shows what can happen when too large a segment of the investing public is ignorant. Not only must capital sources be varied, and information good, but knowledgeable investors, who still learn largely through experience, must be able to make wise choices. That wisdom is made up not only of economic know-how but of judgment based in character.

A large part of investment capital in Europe and North America is invested by relatively well informed professionals, through pension funds, insurance companies, and mutual funds. I stress "relatively," as

---

9. It is a form of *Seinsgeschick,* a gift of epochal being, in Heidegger's terms.

10. It is not that the Colombian élite does not understand this; it is rather that, for historical reasons, the concentration of capital in their hands exists and they have no personal incentive to loosen their control. The limbs on which the society needs to move forward develop strength very slowly. I was amused when working in Bogota a dozen years ago at the élite's eagerness to show off THE self-made man, the successful shopping center developer Pedro-Miguel Gomez. One soon began to wonder, "Is there a second one?" Given the present vigor of Colombian development, I assume that by now there is.

many "analysts" turn out to be quite superficially informed. Also in an "information world," it turns out to be quite difficult to find out *what is really going on.* I know this as a corporate director who has been "mushroomed" (kept in the dark and covered with . . . fertilizer) by his own CEO during a period of half a year when things were going badly and he was in the throes of a bout of *denial.*[11] Victory of sinful heart over smart *caput.*

Cooperation in Investment, Innovation, and in Just Making the Machine Run

While modern man has tended to conceive of himself as a self-standing individual and admires the risk taker, enterprise from beginning to end is social, requiring cooperation.[12] Even the lone investor is dependent on others for information. The manager is like a symphony conductor, manipulating the individual needs of the players to draw from them the particular element of cooperation and of creative innovation his enterprise requires.

One of those interests may well be the basic human need to build up a secure relationship, to become genuinely part of the operation, to be able to count on a position, so long as the situation remains healthy. The boss and his chief engineer and his accountant may develop good relations, they know they can count on excellent performance, and the subordinates come to feel that the boss values their contribution, considering them not just personnel to be exploited but really an essential part of the team. That is indeed a valuable feeling, worth sacrificing for. One of the secrets of the success of the HTX is that in making this vast machine run there is much that is just plain *fun.* Exercising enterprise, daring, skill, inventiveness, a joyful, fruitful cooperation, excitement of discovery, managing to get people to do what your vision demands is fun, and many fabulous machines are really not boring.

I recently read in the Canadian Pacific employees' paper about the

---

11. I also saw at close hand a CEO of a $2 billion trust company that got into terminal trouble mushroomed by his own board chairman (and principal owner)—imagine, the president did not know essential aspects of what was going on during the terminal crisis! Moral factors are central in the world of information—there are often great efforts made to distort information or keep it from people.

12. Mario Laserna Pinon told me that when he studied in the middle 1940s at Columbia University he was struck by the contrast between American forms of cooperation and the extreme Enlightenment individualism that he saw as devastating to Colombia's development. Upon returning to Bogota, this recent college graduate founded the Universidad de los Andes, "to alter the Colombian mentality." That private university, the most successful in that part of Latin America, has formed five prime ministers and dozens of ministers.

nine days of around-the-clock backbreaking work it took to rerail twelve 113–ton cars of potash that had derailed in the upper spiral tunnel on Kicking Horse Pass, British Columbia, blocking the mainline and leaving no room for heavy cranes. So the cars had to be unloaded with clumsy, slow aspirators and shoveled clean, then jacked up, and track rebuilt under them by hand. The writer, apparently genuinely impressed by the workers' dedication, waxed eloquent on "the heroic devotion of supervisors and car men and track workers." Do they love the company? Despite what I said above about railroaders' attitudes in my day, probably more than they would ever admit. (Railroading has its romantic dimensions, and the Canadian Pacific Railway—"the National Dream"—is wrapped in mystique). Were they perhaps motivated by a good masculine trait—"a tough job to be done, let's show what we can do!" For sure. But superb cooperation was evident, qualities of *caput*—skill—and of heart. There can be great poetry in high technology. Grinding downhill through the two spiral tunnels with 14,000 tons of potash shoving at your four 4200hp diesels, which whine as their dynamic brakes and pumping air struggle to keep this mass under control, makes the time go by fast for locomotive engineers. Office-bound types often forget this side of the HTX and are reduced to getting their kicks out of winning little political maneuvers—the great art of bureaucratic fun and games—or pulling off a good speculative coup in the markets.

But consider this comment from a HTXRG techie who works in multimedia:

> You should watch a multimedia office work: we technonerds have the same sense of exhilaration when we finally get through debugging a program! What is interesting about your metaphor of the train is that it is still quite physical: the feeling of that huge weight, the speed, the tunnel, etc. In the HTX world as I know it, there is increasing exhilaration over pure symbol manipulation. Think of the guy who shut down that London bank (Barings, March 1995): he claims that he is really a scapegoat, and that he only lost a couple hundred million dollars! People are leading battles, having arguments, getting high, on pure symbolic interchange. If you look at computer dating, it is almost like this, an almost ethereal experience where one is completely represented by flashing text on the screen.[13]

One aspect of this cooperation I have puzzled over is the significance of *anonymity* in most HTX inventiveness. Take the recent development

13. C. Woodill, personal communication.

of the Boeing 777, the first American passenger jet entirely designed by CAD-CAM. Over one million parts had to be designed by 7,000 engineers to fit together in the prototype. First, I cannot fathom the skill of industrial systems engineers in parceling out and bringing together successfully a million separate tasks of innovation. But how do the individual engineers feel when their particular blasts of astuteness and pure being-inspiration get anonymously melded together (well, not altogether anonymously, Ted Schmitz can still know that his weight-reducing breakthrough in the design of the aileron electric controls contributed eliminated 398 grams of weight, and every little bit helps!)? Collectively, all certainly take pride in "their" airplane, 7,000 engineers' breasts swell on "roll out"; it is "a team win." There is something touching in this anonymity, but I feel there is more to it than that, and I am at a loss to put my finger on it.[14] In any event, one of the most challenging managerial tasks in the HTX is both inspiring the research and development people and then getting them to contribute sufficiently to the bottom line.

Reflecting on this conundrum, one of the HTXRG made this sobering comment, reminding us that, for all its attractiveness, at its limits the HTX may be destroying community:

Is Boeing any different than the Church? Think of how you go to church, and how you are at several different levels of community at the same time: you go as an individual worshipper with your private prayers, you go as a family unit, you might be part of a church committee (maybe 10 to 15 people), and you also feel part of the entire Church. If you are part of a denomination, e.g., Catholic, Muslim, Jewish, etc., then you feel part of a worldwide community. Is there really a difference between that and the engineer? It seems to me one of the things that changes is the efficiency of communication: a hundred years ago, you would not have been able to talk to 100 people in a year, where now you might at least contact that many in a week with fax, e-mail, telephone, media, etc. The whole idea of community, in what I think is your sense of it, is, I believe, an unHTX phenomenon. It is community that is being destroyed: think of the University of Toronto, where there are 80,000 people, and yet nobody seems to know anyone or have any sense of connection. If you go to the Internet level, there is no sense of being part of the Net "community":

14. One engineer pointed out to me that this anonymity can be a protection, allowing you to continue doing what you like doing. When you stand out too much, they pull you up into management.

there might be 30 million people there, but they are completely non-existent unless you get a message from them.[15]

If the human soul is inherently communitarian, how far can allegedly anticommunitarian HTX tendencies run roughshod over what is a human necessity? Is the fact that the small enterprise can retain an element of community, based on face-to-face interactions, not one of its strengths? (I shall address the "so what" of this in the last chapter.)

Effects of Government Policy on Two Basic Areas: Capital Flows and Education

Capital Flows: Are Governments Suffocating the Leading HTX Countries?

These human considerations at the core of the HTX are not the whole question, however. Turning to two other basics, capital flows and education, we must consider phenomena on a scale polar opposite to small-scale community: the governmental.

Government policies play a large role in how capital is invested and in the cost of doing business. Some of what governments do in this area can be helpful, indeed necessary, such as fighting fraud in capital markets and struggling to maintain an even playing field through anti-monopoly legislation. And very few would argue there is no need for a state-provided social welfare safety net, which directs capital into housing and hospitals; inefficient though the social help schemes generally are, they cannot reasonably be dispensed with. Governments are today also the major providers of transportation infrastructure, but whether they are the most efficient is being questioned, and some heretofore unimaginable privitizations—Japan Railways and British Rail for instance, and even Air France!—are now being tried.

The danger of the Bloating State crushing the HTX economy is now widely acknowledged. Symptom: U.S. annualized weekly wages in Constant 1992 wages peaked in 1972 and have fallen $4,916 from their high, and this at a time of unprecedented growth in government regulation.[16] It would be foolish of course to blame this stagnation uniquely on the inefficiencies of government. The oil shock, for instance, played a role, as it constituted a massive shift in capital, which then had to

15. Comment from C. Woodill. Of course the Internet, no more than the phone system, is no community, although on both, interpersonal relationships of sorts can, with difficulty, be kept up.

16. *Strategic Investments,* Canadian Edition, March 16, 1994, pp. 1, 10.

trickle back into the world economy, just as inflation eroded the retail price back to preshock figures. Still there is some corroboration: the stagnation of key, highly socialist Scandinavian economies and of the Canadian as well.[17] Second symptom: Despite eight years of a Reagan administration intent on reducing the size of government, bureaucracy grew at both the state and federal levels.[18] In Canada, five out of seven people live basically off of government money; how much can courageous politicians count on the electorate to back them when they set about to throw hundreds of thousands of well-organized bureaucrats, teachers, and social workers out on the street?

Is this a structurally irreversible situation in the democracies, which will change fundamentally only after an economic collapse? At the moment we are seeing signs that the electorate is willing to tolerate the pain of "downsizing" in order to reduce intolerable government budget deficits. In many American states constitutional limits have been put on governments' rights to tax or to spend money they do not have. Nervousness exists everywhere governments are absorbing more than half of income.

The recent difficulties of the central command societies with their managed markets are rife with implications about HTX ekistics. Intergoverment efforts to remove obstacles to international trade, the expansion of intergovernmental negotiations, and the new international economic organizations have contributed to a three- or four-fold increase in the portion of the economies of all First World countries absorbed in international trade. This has added to the wealth of many nations, but at the same time some of the resulting distortions are generating serious tensions, especially the trade imbalances between the U.S. and Canada, on the one hand, and Japan, Taiwan, China, and South Korea, on the other. High "structural" unemployment is a problem in both Europe and Canada.

Here we discover a typical HTX-style dilemma: the need has obviously arisen for a new level of sophistication in understanding the civilization, economy and government of the Other, and this on a

17. And after three years of Conservative government, the Swedish people, having become dependent on their social welfare system, voted the Social Democrats back into office!

18. Granted, the Republicans never once in the last forty years controlled both houses of Congress. Now that they do, but not the executive branch, it will be interesting! My bet: after enormous huffing and puffing, only a slight slimming of the size of government.

planetary scale, as well as the need to understand the "give" in one's own systems; and then negotiators must learn together how to play these complex systems off against one another. The challenge is horrendous, yet the negotiating personnel shift constantly.

Perhaps the very fluidity and complexity of the HTX keeps these tensions from becoming a hardened crash—there seems always to be room for another round, so keep the gunboats at home.[19] It is more like a many-sided taffy pulling contest. And all pulling at the taffy are many governments and many layers of each government, along with large corporations playing simultaneously on numerous fronts, for this taffy is made up, so to say, of independently operating ingredients.

Careful contrast of the present situation with that in Europe in 1913-1914, where the taffy pulled apart, might prove instructive. The danger now may be most imminent with rogue states, caught in their little corners, insufficiently HTX and thus inflexible, both in their narrower inner structures and in their boxed-in situations, and so tempted to lash out in frustration. When they are strategically located, like Iraq, Iran, or North Korea, the big consumer states are obliged to react to their tantrums, but it is getting terribly expensive to do so. And as more of them get the bomb, the game becomes quite interesting.

On the immense danger posed to the world by the Grand Rogue, China, the corrupt, ideologically bankrupt, totalitarian dinosaur with nuclear capability and quirky control over a fifth of mankind, I shall not here comment further. The Japanese, understandably, are obsessed by this massive reality on their doorstep. A Pakistani engineer who worked five years in Japan told me that the Japanese give themselves fifty years to come to terms with China. If that implies they have fifty years to get ready for a great confrontation, then, I believe, the Japanese are hopelessly optimistic.

Education

The other pillar of the HTX economy, education, is also heavily dependent on government. Education is more than honing skills. Underlying whole ranges of skills is the need for certain basic mind-sets and thinking abilities that make the HTX possible, and that only a certain kind of general education can provide. Way beyond the basic liter-

---

19. Is the Yugoslav disaster a result of the region being insufficiently HTX in mentality, the clash coming as a result of deeply rooted ethnic loyalties, with little sense of the give-and-take negotiation that is more HTX?

acy primary education ought to provide, necessary to advance beyond primitive agriculture, comes solid competence in math and science and the ability to write well and to read complex texts. As though all this intellectual and skill-forming education, with its numerous subsets, were not enough, somehow, in the process of all that educating, the character and the heart have to be matured. That is a task formerly accomplished largely by intact, extended families, but also by tough, "illiberal" schoolmasters and -marms, devoted nuns and teaching brothers, village parsons, and tough taskmasters in the office and shop. Now, with half the nuclear families split up and family members living thousands of kilometers apart, and a crisis in religious vocations, with many offices and factories at a scale breeding impersonality, we leave both sets of tasks more and more to corps of well-paid unionized schoolteachers, quirky, tenure-spoiled university professors, and narrow company training courses and even "company universities." In a school atmosphere affected by the students' competition for jobs, many pick up the message that "success" is assured by beating out the other guy—me first and last, which is not helpful for fostering the fruits of cooperation.

No modern state is, or ever will be, up to this overwhelming educational task. Despite the fact that two of the largest categories of government expenditure are education and medicine, with private companies also spending vast sums for employee education and their own technical institutes, there is nowhere in the HTX adequate thinking out of educational strategy and effective coordination of educational effort with the seemingly infinite variety of needs for different forms of general education and specialized skills.

When at the height of the Cold War I would get depressed about inadequacies of the NATO forces, I would console myself by remembering that the Warsaw Pact forces, while more numerous in some categories, were even worse trained. It is like that with education: If we consider the competition of the most advanced HTX countries, and how each stands as regards education, the reality is too complex to form a valid overall opinion. Not only do the needs for specialized skills change too rapidly for academic bureaucracies to keep up, but in the area of more general education ideological confusion is rampant.[20]

20. At the depths of the last recession, with unemployment approaching 12 percent, there were 450,000 high-skill jobs going begging in Canada, despite the fact that for the last two decades the need for better job training has been a major and perennial political issue and despite the expenditure of hundreds of millions of dollars annually on "job retraining," usually ill-fitted to actual needs.

Basically, I believe it fair to say HTX education is in a mess. Everywhere. In different kinds of messes, and we hardly understand what
the various messes signify!

Again, does it really matter all that much to the overall well-being of
the HTX? One might answer: So long as both educated and skilled
people can move around freely, those with certain strengths, due to
personal circumstances and the peculiar abilities of certain schools or
systems, will move where the market dictates. So the issue is transformed into a question of social and geographical mobility, and, well,
too bad for the families that are torn up; the extended family *has had it*
in the HTX anyway.

A friend commented:

> I remember being in Enrichment Class (a great euphemism in itself—
> the only other things that are enriched are metals, flour, and cereals)
> and finding myself in a class of over-competitive HTXers, who would
> constantly compete for grades, but would not be able to talk to their
> parents, or go on a date, or get a job. I think though, that the world is
> quickly changing, and the universities are going to start finding them-
> selves out of a job as the markets themselves take over educating.
> Even today, you can now go to Microsoft school, and with no degree
> at all, get a job as a junior programmer at Microsoft. In the technology
> fields, people are more and more looking at work experience and raw
> talent over a liberal BA or even a general BSc.[21]

There is much truth in this. My worry lies most with ignoring needed
foundations in general education. A young cardiology professor who
has been learning to do research in a molecular biology laboratory
complained to me of the researchers' lack of elemental logical analysis. She agreed with my instant diagnosis: Poor culture at home, followed by inadequate primary and secondary education, which now
undermines them at a basic level of *thinking* and communication.

Obviously, beyond a pretty imbecile level, thinking and communications are important for passing on highly focused, specialized skills.
If the student has poor math training or little grasp of the basic sciences, if he has trouble analyzing complex objects, he is not likely to
turn into a very skilled technician, let alone an engineer worthy of the

---

21. C. Woodill, private communication. Bruce Stewart when he was a CIO pre-
ferred bright high school graduates to B.S. holders from the university computer
science program, for the latter had to unlearn, so fast is technology moving.

name.[22] If he is inarticulate, incapable of writing a good report, or defensive with people, because his human base is narrow, he will not become a brilliant manager.

Widespread poor-quality thinking throughout basic sectors of an HTX society will lead to a general lowering of the quality of life. Serious societal discontent can build as a result not only of poor character but of just plain confused thinking throughout the systems.

When such a dangerous situation appears severe enough, can a given HTX country simply *manage* its way out of the mess? In North America, as regards general education, and in Europe and Japan, as regards ingenuity and competitive spirit, the alarm bells have been ringing for some time.

The Internationalization of Mentalities

An international HTX class of managers, scientists, technologists, diplomats, and entertainers is growing, with many specialized worlds of people able "to speak one another's language." Think of the world of pilots, for instance, or "the oil men," "the guys in the pinstripes of high finance," the big-time entertainers, each with their internationally recognizable costumes—the pilots' uniforms, the hard hats and dirty coveralls of oil field workers, the Wall Street traders' somber suits, and the long hair, or shaved heads, and exhibitionist clothes of the popular music stars.

Still, in even the most HTXed countries, the "internationals," with their supernational (but still narrowly specialized) horizons, constitute a small part of the local population. Moreover, one wonders just how far into the mentality of the masses their influence really extends and, indeed, how profoundly they themselves have been transformed. The entertainers reach into the psyches of the masses, but there is debate everywhere about the nature and extent of their influence. I have generally not been too impressed with the depth of the insight into other cultures and other peoples of many of the internationals I have known. Their contacts with a small elite among other peoples remains highly instrumental, and they rarely have either time or inclination for contemplation of the difference in cultures. There are exceptions, but I

22. My daughter tells me that this is indeed the case with the most competent, intelligent operators on the super-sophisticated lines at her IBM plant: they bump up against the limits of their education, and all the goodwill and general brightness in the world will not substitute for the more general knowledge and the highly developed mind-sets of the well-educated engineer.

have been more impressed with how isolated from the depths of the real world many internationals remain, looking down from their managerial, scientific, or technical stratosphere, and their Hilton Hotel suites. The exceptions are the kind of people who would have made good missionaries: they are interested in other people as *souls*. Instead many HTX internationals may be more like our concept of the bad missionary, the "imperialist" who forces his worldview upon others. Their extensive foreign experience may more remove them from living contact with their own culture than it immerses them in any other.

Permit me a moment to sketch a standout exception, paradigm of the old-style international businessman: He started his career working for Baron Detterding, the man who shaped the Royal Dutch and subsequently served seventeen years, starting at the age of thirty-eight, as director general of *Shell française*. At twenty-five, he was sent around the world for a year by an oil millionaire and father of a friend, with the sole assignment of looking for investment opportunities. André Goldé and my friend Jacques Vidal spent, for instance, an entire month in Japan. Vidal was inspired by that contact to become a lifelong student of Buddhism, and by the time he retired he had developed the Gestalt of an English gentleman. He had escaped the infamous Jew hunter Klaus Barbie (he was reportedly number one on his list in Lyon!) by walking at night over the Pyrenees (carrying in his pocket a little copy of Aeschelus' *Orestia* in Greek with no dictionary so he could read while interned in Spain), to spend the rest of the war helping the oil intelligence office in London. He possessed a very personal, much reflected-upon worldview and a profound appreciation for the peoples of different cultures with whom he had worked. This remarkably cultivated man of the arts was a genial manager and a perfect gentleman. He was no dupe of HTX hype. During his years as director general, he devoted the hours from 11 P.M. until 2 A.M. to his personal cultivation. He was a "people person," who devoted three quarters of his time, he told me, looking for future management talent and personally training these young men. I wonder if today's HTX whirlwind can still produce a Jacques Vidal.

I was discussing the question of degree of HTX penetration in the managerial class with a successful American engineer who is in middle management at a large Canadian public utility. A very careful planner in his own life, he opined, optimistically: "Perhaps people will take from the HTX just what they want, 'learning the language' and adopting the necessary attitudes just to the point of achieving their particular goals, but no farther, and adapting these instruments to fit their own desires

and their own peculiar situations." So an ex-samurai class of senior pol-
icymakers in Japan will do the necessary to keep their companies internationally competitive, profiting, for instance, from closed national
markets, where consumers are forced to pay high prices to accord the
companies the economies of scale necessary to offer unbeatable prices
abroad. Then, in their private lives and their ways of relating humanly
with subordinates and with family, they will continue ancient styles of
cultural and family life. This contrasts with the rougher style and more
short-term management of Hong Kong entrepreneurs in a personal
survival mode, building from small enterprises on a base of (then)
cheap labor willing to work unstintingly to get a beachhead for life. But
these too will depend on the extended Confucian family, ferocious family loyalty, and family "capitalism," lending meager savings as start-up
and expansion capital.[23]

A Canadian friend who has built the largest new factory in Russia (to
produce candy bars!) is of the opinion that it will take a minimum of two
generations to begin to produce a corps of Russian managers with an
entrepreneurial form of HTX mentality. Mostly, HTX mind-sets will be
passed on through implicit contact; the rest will come from sending off
cadres to get MBAs. My friend complains, for instance, of a total lack of
awareness among the Russians that time is important. He worries that
his company's just-in-time inventory policy cannot work when suppliers
have no sense of urgency. He sees no shift in the vast and ancient bureaucracy, in which each bureaucrat demonstrates his power by ob-
structing. While partially HTXed, the being of Russia is different from
the being of Europe and of America, remaining caught between the extreme socialist mold and the chaos of "savage capitalism," with almost
nothing of a modern Western sense of the rule of law.[24] Meanwhile bureaucratic Soviet-style formal mass education continues to grind along.

23. A Japanese guru laments the unwillingness of Japanese companies to en-
trust management of foreign operations to locals. *Business Week,* 1994 special
bonus issue: "21st Century Capitalism," December 1993, 90. See also the view of
Shojiro Asai, general manager of Hitachi's advanced research lab, p. 103.

24. My friend is painstakingly teaching his own people the importance of deadlines, but he says very few have really interiorized this. The capitalist sense that
"time is money and money must pay over time (interest)" is entirely absent. Only a
few youngsters, he says, are beginning to understand you must produce something
if you are going to justify your "cushy" job. (I wonder how many Canadians really
understand that.) Meanwhile, at startup, the milk needed will come in powder form
from New Zealand. Despite the company's exhortation, "Buy Russia," the Kolkhoze
system has broken down and to date can guarantee no sure supply of milk.

Is "Information" Everything? What about Insight, Appreciation, and Judgment?

The Flood of Information

Because of nature's intractability and because of cultural drag, we know everything is not going to be changed overnight—whatever the HTX hype—and some things *never* will so long as man exists. To a great extent, it is the same old child of Adam and Eve roaring down the information superhighway straight into brave new worlds, a child who is also a "son of God," who claims to be touched by the transcendent. All this adds up, then, to the truth that *education has to be of distinct kinds taking place on several levels moving at different paces,* affecting not only different persons differently, but the same person in different ways, setting up tensions within the person.

So, amidst celebrations of the "Information Revolution" one is well advised to remember that "not by info alone doth man live." (Come to think of it, the biblical quote is, "Not by bread alone doth man live, but *by every word* that comes from the mouth of the Lord," which is an affirmation that, finally, while information *is* central, only *Truth,*—a "word," not just *data*—is useful. A word is spoken, sincerely or insincerely, correctly or incorrectly, by someone and has to be *received* ["Obedience," recall, comes from the root *audire,* to hear] and responded to [*spondeo* = I commit] by *someone:* communication is interpersonal, requiring not just mind but will and character.

Now there is an educational challenge in the midst of the information avalanche! We have already seen Japanese business gurus awakening to it. They ask, "Does anyone really *hear* anything anymore?" *Why* are we so *hard of hearing*? Teachers in Toronto tell me they are increasingly frantic about the unwillingness of students to listen, and worse yet, to communicate. ADD, "attention deficit disorder," is now a recognized psychic disorder. It is exactly what Marshall McLuhan insisted TV would cause.

The Moral Foundation of Sound Information

Bread (= taking care of our material nature) does indeed remain fundamental. But I will not head out for the bakery until either I feel hungry or I remember that I shall certainly become hungry and so had better lay in groceries. This requires more than recognition of the significance of the present pangs in my stomach, or remembering that such pangs reappear with predictable regularity: a further *judgment* is made that it is just now more important to respond to this information than to continue to write this book. That judgment results from balancing factors of relative desire, time, possibility (e.g., of finding something to eat). Behind all those judgments which turn data into information (disinformation if my judgments are poorly founded) is always the principle: *what do I really want,* with the not-so-hidden agenda of balancing the "hierarchy of needs," as Maslow calls them. (*Pace* Saint Augustine, one might put it, "Show me how the man prioritizes and balances all his needs and I'll show you the man.")

A good part of wisdom is getting the balance right. *Truth is indeed symphonic.* But just balancing needs is not enough, those needs have to be regulated by reality external to us; there is a fundamental need for the person with the needs to be in clear-sighted contact with the fullest reality he can access. Data becomes information, and information, truth on the basis of the *genuineness* of what one loves, that means how one's loves are in contact with reality, not projections of fantasy, eager for sound data, not in partial denial.

Often more data can improve the information, and better information should lead to a better judgment about what to do, (including what missing data is still needed and where to search). But much more is involved in making a good judgment than just the necessary information: for balance to be achieved, character, discipline, and hard work are necessary, all of which are fruits of good education, well responded to. After all, not all information is *true,* just as not all data is reliable. Disinformation has a way of obfuscating and twisting data. In the real world, we operate under time constraints, many judgments about complex, ambiguous situations are inevitably "snap judgments," some of them decisions of grave importance, where we simply have no choice but to go now with the information at hand. That requires, when one has to make "seat of the pants" decisions, wise experience to fall back on, and habits of making quick assessments of good quality. (Aristotle grouped these and similar skills under the virtue *sophrosyne,* [note the same root as *sophia,* which today is translated as "wisdom"] which the Romans translated *prudentia,* and which St. Thomas called "*recta ratio agendi*—right reason in acting.") Experience is not acquired in schools.

An Information Flood at the Speed of Light, and the Explosion of Images

During the period when the printing press began permitting wide distribution of data, at the pace of horse drawn vehicles, new institutions sprang up: academies of science, with agreed-upon standardized symbol systems, public schools, and representative governments. But when the telegraph began whizzing data around, *space was annihilated.*

One day after Samuel Morse's demonstration from Washington to Baltimore, the *Baltimore Patriot* informed its readers of a congressional debate over the Oregon issue, concluding the article with this prophetic declaration: "We are enabled to give our readers information from Washington up to two o'clock. This is indeed the annihilation of space."[25] Postman comments on this revolution, "Within two years of this announcement, the fortunes of newspapers came to depend not on the quality or the utility of the news they provided but on how much, from what distances, and at what speed."[26]

At almost the same time as this explosion of news of dubious quality came the photograph, and with it came what Daniel Boorstin calls the "graphic revolution." Postman describes it: "the massive intrusion of images into the symbolic environment: photographs, prints, posters, drawings, advertisements . . . not just supplementing, but tending to replace [print] as our dominant means for construing, understanding and testing reality."[27] We have barely begun to think about educating the young to deal with this "massive intrusion of images."

Postman calls for a "new definition of information," especially in the face of "information [I would say "data"] that rejected the necessity of interconnectedness, [which] proceeded without context, argued for instancy against historical continuity, and offered fascination [rather than contemplation and appreciation] in place of complexity and coherence. Like the Sorcerer's Apprentice, we are awash in information."[28] To this we can add, with some five hundred TV channels coming from "Death Star," interactive TV, conference calls via two-way TV, and shopping via the Internet, "You ain't seen nothing yet!" A whole new level of education is demanded to manage this onslaught.

Controlling the Data Flood: Bureaucracy, Expertise, and Invisible Technologies

This flood of data overwhelms our social institutions. Postman points out "social institutions are concerned with the *meaning* of information and

25. Cited by Postman, *Technopoly,* 68.
26. Ibid., 68.
27. Ibid.
28. Look at the flood of symbols now made available, adding broadcasting and computers to the revolution: In the U.S. today, according to Postman, there are 260,000 billboards, 11,250 newspapers, 11,556 periodicals, 27,000 video outlets, 500 million radios, 98 percent of homes have a TV, 40,000 new book titles a year, 41 million photos taken daily, and 60 billion pieces of junk mail delivered yearly. Now from my computer an endless cascade of information floods onto my desk. Postman, *Technopoly,* 69.

can be quite rigorous in enforcing standards of admission" (e.g., a court of law, and universities, through exams and carefully vetted research).[29]

Because the flood has carried away the theories upon which schools, families, political parties, religion, and nationhood itself are based, American Technopoly must rely on technical methods to control the flow of information. Postman singles out three of these:

First, *bureaucracy*, "an attempt to rationalize the flow of information, to make its use efficient to the highest degree by eliminating information that diverts attention from the problem at hand."[30] Not per se a social institution but more an efficiency mind-set, bureaucracies ignore as irrelevant all information not contributing to its notion of efficiency. "That is why John Stuart Mill called it a 'tyranny' and C. S. Lewis identified it with hell."[31]

Created initially to serve social institutions, bureaucracies, as they grew along with the flood of information to be managed, added to the flood themselves, so that more bureaucracies were created to manage the bureaucracies. From being the servants of social institutions they became the masters and began themselves to address moral, social, and political issues. Take, for example, the huge bureaucracy of the ministry of health in our province. With very few medical doctors aboard, they are now making sweeping decisions in principle and in detail about health care, affecting the well-being of millions of Ontarians.

Postman reminds us that the French word *bureau* first meant the cloth covering a reckoning table; from table, to room, to office building, it keeps that sense: the bureaucrat is interested in the implications of a decision only to the extent it affects his "table."[32] The Eichmann answer is common: I have no responsibility for the policies that drive the place, but for seeing that my little part of it operates efficiently. "Eichmann was an expert, and expertise is a second important technical means by which Technopoly strives furiously to control information."[33]

Second, *expertise*, Postman points to three reasons for the growth of "experts-as-ignorameses," pontificating on every moral, social and political aspect of life, without being expected to know anything beyond

29. Ibid., 73.
30. Ibid., 84
31. Ibid., 85
32. Ibid., 86.
33. Postman, *Technopoly,* 87.

their narrow area: 1) the growth of bureaucracies; 2) the weakening of traditional social institutions; 3) above all, the torrent of information. Expertise

> works fairly well where only a technical solution is required and there is no conflict with human purposes, e.g. in space rocketry or in construction of a sewer. It works less well in situations where technical requirements may conflict with human purposes, as in medicine and architecture. And it is disastrous when applied to situations that cannot be solved by technical means and where efficiency is usually irrelevant, such as in education, law, family life, and problems of personal maladjustment. There can be no experts in child-rearing and love-making and friend-making.[34]

Third, *invisible technologies,* are essential to the expert for handling the information flood: the "softer" technological machinery such as IQ tests, SATs, standardized forms, taxonomies, and opinion polls. These technologies often go unnoticed in their role of "reducing the types and quantity of information admitted to a system," and hence their role of reducing traditional concepts is missed.[35] *"There is for instance no test that can measure a person's intelligence."* But an IQ test "transforms an abstract and multifaceted meaning into a technical and exact term that leaves out everything of importance." We come to believe that this score is our intelligence, as we believe polls, as though what people really believe can be encapsulated in such sentences as "I approve." (The manipulation of polls for the most earnest business of the republic has never been more obvious than in the Clinton-Lewinsky affair.)

> When Catholic priests use wine, wafers and incantations to embody spiritual ideas, they acknowledge the mystery and metaphor being used. But experts of Technopoly acknowledge no such overtones or nuances when they use forms, standardized tests, polls, and other machinery to give technical reality to ideas about intelligence, creativity,

34. Ibid., 88
35. Ibid., 89. Baudrillard's (*Simulations*) example comes to mind: By making you choose either Coke or Pepsi, both companies ultimately win, because you can't choose root beer, orange juice, water, or nothing at all. In general, all pop works this way, because each owns a brand: Pepsi owns 7up, Crush, and Diet Pepsi; Coke owns Sprite, Minute Maid, and Diet Coke. Even with the upsetting that companies like COTT have done, this is still largely the case.

sensitivity, emotional imbalance, social deviance, or political opinion. They would have us believe that technology can plainly reveal the true nature of some human condition or belief because the score, statistic, or taxonomy has given it technical form.[36]

Postman admits the utility of such "information control" but wisely adds, "unless administrative decisions [based on such technicalized information] are made with profound skepticism—that is, acknowledged as being made for administrative convenience—they are delusionary."

> In Technopoly, all experts are invested with the charism of priestliness. Some of our priest-experts are called psychiatrists, some psychologists, some sociologists, some statisticians. The god they serve does not speak of righteousness or goodness or mercy or grace. Their god speaks of efficiency, precision, objectivity. And that is why such concepts as sin and evil disappear in Technopoly . . . sin becomes "social deviance," which is a statistical concept, and the priests call evil "psychopathology," which is a medical concept.[37]

Postman is here brilliantly highlighting something fundamental about the HTX being. He agrees with David Bolter, author of *Turing's Man,* that the computer is the *dominant metaphor* of our age, defining it by suggesting a new relationship to information, to work, to power, and to nature itself, redefining humans as "information processors" and nature itself as "information to be processed."[38] He is right in condemning Technopoly's tendency to think all problems require "technical solutions, through fast access to information otherwise un-

36. Postman, *Technopoly,* 90. See J. Weizenbaum, *Computer Power and Human Reason: From Judgment to Calculation* (San Francisco: W. H. Freeman, 1976). He is one of the great pioneers in artificial intelligence and critical of exaggerated claims made on its behalf.

37. Revolt against such professionalism brought into existence the independent living movement. It is a civil rights/consumer/self-help movement that was started in California by a few severely physically disabled men who were living in residence. The independent living movement's basic thesis is that the consumer of services knows the most about his own needs, so that if someone has a disability, the disabled person, rather than the doctor, is the expert. What has happened in recent years is a transformation of the mentality in disability circles: persons with disabilities were seen as people who needed to be cared for out of charity; they now see themselves as citizens with rights and the expertise that comes from living with their disabilities. The view evolved out of being occupied by HTX doctors, psychologists, and governments.

38. Postman, *Technopoly,* 111. See J. David Bolter, *Turing's Man: Western Culture in the Computer Age* (Chapel Hill: University of North Carolina Press, 1984).

available."[39] "Our most serious problems are not technical . . . If families break up, children are mistreated, crime terrorizes a city, education is impotent, it does not happen because of inadequate information."

There is truth in Postman's insistence that all societies, through their social institutions, have sought to control the flow of information. But Postman's opinion that earlier forms were, if not more benign, at least somehow a bit less frightening because the quantity and quality of information to be controlled were far less, makes me a bit nervous. That may be true, too, but a lack of deeper criticism by Postman of the whole idea that man is meant to be basically a controller leaves this discussion floating. Such control may be Confucian, but not Buddhist. It may jibe with Roman practical philosophy, but not with the noblest Stoicism. It is foreign to Islam—God, not man, controls, although great *Kaliphi* may have been in practice superb controllers; and it is square against both Talmudic and Christian anthropology, although extremes of "legalism," which are constant temptations in both traditions, and Jews handing Christians to the Roman authorities and Spaniards handing *muronos* to the Holy Inquisition are flagrant contradictions of the respective traditions.

Control, control, hmmm. The Stoic and the Buddhist may use techniques to bring themselves under control to the point they can at last afford to let loose, but the Christian recognizes in his utter dependence on grace that no technique can do anything more than help one to hold open amidst the HTX buzz a space into which the gifts of being—*esse* first, primarily through creation, then, with the coming of man, being in the sense of the light of meaning—can pour.[40] Yes, we are, every day of our lives, perhaps inevitably, controllers; most human knowledge aims at control. No, we are not meant to stop there, ultimately we

39. Postman, *Technopoly,* 119.
40. The foundational gift, Jews, Christians, and Muslims, in believing revelation, agree is creation, the whole dynamic unfolding of the complexifying universe, about the details of which we are learning more but still know very sketchily. That is a great gift of unfolding *esse*—real existing being. The Christian also believes the coming into history of the divine *Logos* in the person of a humble Jew, Jesus of Nazareth, was a real event—again *esse,* as were his deeds and pronouncements. His words, like the imagemaking of all who add meaning to human life, are gifts in the order of being as world-founding interpretation, which have as condition for their possibility gifts of *esse,* especially human nature, and when genuine, that is when well rooted in the reality of *esse,* will, through human work, further affect existence, in the form of cultural objects, including those intersubjective social realities, rooted in meaning and habits, institutions. The Church is the fruit of Jesus Christ's work, continued by obedient apostles.

cannot control our destinies, we are meant to contemplate and appreciate, as Postman sees very well, not just exploit nature and take God's grace for granted, and we ought to avoid damaging the freedom of others. So we are not supposed to be first and foremost controllers but rather neophyte lovers whose controlling simply reveals vast remains of insecurity. Especially in HTX phase two, our educational schemes had better start being mindful of that.[41]

A Deeper Look at the Nature of "Information"

As "information" constitutes the brain waves of the HTX, I believe we should probe even deeper into its reality. Recent HTX developments help us to understand it better than before. Of course, I do not apologize for calling such an "ethereal" phenomenon a reality, and I'll go further: information is an essential dimension of being. To adapt a famous dictum of Heidegger's: Being never comes to be without information, and information comes to us out of being.[42] In so doing, information illumined by, and added to, being provides *the contexts and the material* for our judgments.

Data, which itself as given is always already interpreted, has to be arranged properly (further interpretation) to be useful, to become *informative*. Every datum is an interpreted form, and this form becomes part of a larger form in being taken up into "in-formation." The process of assuring its position in a configuration or Gestalt can reveal both the agenda behind our desiring a particular arrangement and serious gaps in the data. This requires, if time allows, further research to fill out the figure, and to begin to deepen and extend (to an enlarged context) the form emerging in our grasp of the configuration.[43] One can minimally register but cannot *attend* to a disordered flood of data. One of the reasons the Three Mile Island nuclear accident turned from a serious but

41. The Christian believes we are meant to be open to God's love, which does not control but frees us. We are called to move from the illusion that we can ultimately control to the reality that we are a gift of creation, that we are loved and are called to respond in kind. More on that when we return in chapter 6 to consider in the HTX setting of the Great Clash of Anthropologies.

42. *Sein nie west ohne Auskunft, and Auskunft kunftet nur aus dem Sein.* This is my adaptation of Heidegger's thought, hence I offer no precise reference.

43. The grasp of a Gestalt (sometimes basically only a sensual perception with a necessary minimum of intellectual insight—e.g., "this is an amorphous ball-like Gestalt with some thickness") is a first step toward insight into the underlying form. St. Thomas reminds us that we remain very much on the surface of most of the things we deal with every day, sensually, practically. Going deeper into the form requires a richer intellectual penetration, in some sense always causal, digging below the surface to understand why the thing is as it is.

manageable incident to a near-catastrophe is that the operators became confused and rattled by the simultaneous sounding of alarms
and flashing of warning lights, an overload of disordered data, set off
by a bad judgment about the first (accurate and manageable) information, when the operators chose to consider as an error an (accurate)
reading that said a valve was stuck.[44] Well-trained nuclear operators,
like airplane pilots, acquire on the simulator the experience to make
sound snap judgments in complex situations, integrating a sudden
burst of (frightening!) data into a complex context that has been rendered familiar by long practice, strengthening the possibility of pru-
dent lightning judgments.

At the base of all ordering of data is always *a question:* In the light of
what you want to know data has to be selected and "generated," someone has to go after it, observing, measuring, ordering, criticizing reliability, reporting, and distributing to points where its ultimate desirable
significance can be judged, all different kinds of interpretation, hierar-
chically related.

On the basis of their study of how information is used in the modern "virtual corporation," Davidow and Malone argue there are four
distinct categories with which every corporation has to be concerned.[45]

—"Content Information": about quantity, location, and types of
items. Essentially historical in nature, it records what an employee has
done, when an individual was born, where inventory has been stored,
what a customer has ordered.

—"Form Information": as complete a description as possible of an
object. "Billions of computer operations are often needed to generate
the huge data bases that describe form" (e.g., the parts of an automo-
bile and how they fit together). Like content information, it describes
what has been or is, with little about the future.[46] Hence the need for
forecasting. This produces

—"Behavior information": To predict the behavior of a physical object a computer must be able to simulate its motion in three dimen-
sional space through numerous discrete steps in time. This takes

44. Such data does have to be critiqued, as false indications are fairly frequent
on systems wired all over with sensors, information systems deliberately redundant.
(My son tells me that one of his company's plants, Darlington Nuclear Generating
Station, [four reactors, nearly 4000 megawatts—the world's most powerful] contains enough wire to stretch end to end around the earth five times.)
45. Davidow and Malone, *Virtual Corporation,* 67–72.
46. Recall *essence* = *Wesen,* the already having been.

massive computer power, often beyond the capacity of even the latest supercomputers. But because real life testing is usually so expensive and even dangerous, computer simulation remains an ideal toward which the virtual reality world is working mightily.

—"Action Information": "As computing power grows inexorably cheaper, compact, and more powerful, it is possible to build machines that not only gather and process information but act upon the results. . . . As industrial robots, they can accept information and use it to shape mechanical parts, inspect and pick and place parts, and, in a scenario right out of science fiction, build the next generation of industrial robots."[47]

I shall now reflect on the *ontological implications* of some of the IP factors involved in all four categories of information.

Accuracy, Coverage, and Proof of Both

Often the accuracy and adequacy of coverage of the data about whatever interests us leaves much to be desired, sometimes for lack of time, sometimes lack of zeal on our part, sometimes lack of grace. In the temporal deployment of being, besides will, especially intensity of interest, there is the grace of what the thing will yield ontically in the time at our disposal, given the limits of tools for seeing, of concepts capable of directing observation in a certain direction ("I just didn't see that!"), and social support for a certain arduous way of inquiring. Beyond that hovers the mystery of what being will allow to be revealed now, or perhaps ever.

As it is always hard to be aware of what you do not yet know in what you do know a bit, *humility* is required. (We do not hear often enough, "I am really not sure; my judgment is based only on vague impressions.") On the other hand, it is a classic mistake to seek a kind of accuracy or certainty simply not feasible for the kind of object you are dealing with. This was one of the gravest widespread errors of the nineteenth century, which has been given the name "positivism."[48] The positivists were enamored of the kind of proof obtainable when a

47. Ibid., 71.

48. The name comes from the classic of Auguste Comte, *Cours de la Philosopie positive*. See E. Gilson, T. Langan, and A. Maurer, *Recent Philosophy* (New York: Random House, 1966), 266–89. The outmoded disciples of this much-refuted way of constricting things still hang on in large numbers in the university. Even someone as brilliant at Stephen Weinberg can still be infected. See his chapter, "Two Cheers for Reductionism" in *Dreams of a Unified Theory* (New York: Pantheon, 1992).

physical reality is analyzed with all variables isolated by being put to repeated experimental test. It is foolish to demand this kind of knowledge of, for instance, free operations of the human spirit or, for that matter, the history of the cosmos, which is a onetime event. Fortunately, the case of Weinberg not withstanding, positivism has lost much of its appeal at the level of particle physics and quantum mechanics and, for many, even in cosmology.[49]

Feedback and Speed

Proving the accuracy of data requires *feedback:* the system has to allow one to return by a slightly different bias to the object, the nature of which has begun to be grasped by insight, to establish the extent to which the data presents relatively stable aspects of the object correctly (or if the object is very dynamic, consistently unfolding aspects). It requires *consistency:* If, now that I believe I know what is the thing or person or situation with which I am concerned, some of the data appears anomalous, it may not be the data that is incorrect but rather the conclusions I have already drawn about the essence of the thing or situation. Similarly for *coverage:* The knower may early on forge for himself a vague overall view of the object and situation. In some instances this may take the form of an explicit model or working hypothesis, but more often in everyday life it is a vague image or a commonsense familiarity. He should then put the data to the "How do you know?" test—what data, of what quality supports this or that aspect of the "overall view"? And what might be hiding in the anomalous data? A dialectic back and forth between holistic level and lower-level contact caught in the data proceeds as one comes to know the essence of the object more surely, and hence deepens his grasp of the form.

Given that the objects we need to know are in flux, our system of knowledge has to allow the best feasible "real time" feedback, essential to both "behavior information" and "action information." We must constantly be altering our "model" in response to changes in the data, seeking to anticipate how the thing or situation will develop, and returning to the data with new questions posed by changes in the overall picture. A certain *speed* is in most instances essential.

Bizarre as it may seem, I cannot recall the issue of timeliness and speed ever entering into an academic discussion of ethics! It does not seem to have occurred to many philosophers that the need *to decide* on

---

49. The history of being has moved on, but after every epochal change, some remain stuck in the old tradition, and for every past epoch one can find throwbacks.

the basis of whatever information you *now* have on hand in an ever-changing situation is one good reason for having someone in command, hence why *obedience* is a practical necessity. I raise these moral considerations here en passant as a reminder that the ultimate decider is a time-bound human being, and his decision is an act of freedom. "Decisions" delegated to machines, even computers, are simply fallout from earlier human decisions (the humans set the parameters within which the machine must discriminate, with no creative freedom remaining) and will always demand further human—that is genuine—decisions: These preset mechanical "decisions" maintain a fixed Gestalt, a set scenario of "becoming," with preset algorithims—allowable options. In the real world such "closed loops" are always, in fact, operating within dynamic, unfolding contexts, and hence require judgmental integration into the englobing dynamic reality.

Speed, Coverage, and Accuracy versus Cost

In most systems, speed, coverage, accuracy, and the quality of feedback can be improved if the master of the system is willing to pay enough. For instance, as perfect reliability in microprocessors is the goal, 80 percent of the cost of manufacturing and packaging the microprocessor (mounting the chip on its complicated substrate) is expended in testing the chips to see that each and every one will perform as promised. That is an expensive information gathering system, indeed, but it is judged cheap compared to the alternative: a certain (unacceptable, as decided by the market) rate of computer failure. Underlying the whole setup is the very fallible, complicated human managerial decision of just how much is enough to spend on assuring reliability.

I am also assured that the vigilance (moral quality) of the operators is vital on the hyperautomated lines in which the many steps of this manufacturing process go forward. They have to be alert at every instant to the flukes in the sophisticated but imperfect machines whose quirks at critical points only machine-aided human perception can pick up—human perception: those great holistic computers in the crania, programmed by experience, and activated by *will* formed by character and good habits! (Try as it will to exorcise man, the HTX system remains "all too human!" The highly trained worker is increasingly recognized as the ultimate resource.)

Speed and cost! Data can be gathered, processed, transmitted and distributed, stored, and retrieved not only in some respects close to the speed of light but at prices for some aspects 40 percent less than a year ago: speed zooming, prices falling, at rates beyond the imagination of man even fifty years ago!

This speedup is exhilarating and should be *thaumatzein*-causing (Aristotle said all philosophy begins in *astonishment.*) I am even more astonished at the lack of astonishment in the average guy faced by this revolution. How fast human beings adjust, and how quick we are to take the greatest wonders for granted! The human imagination cannot cope with speed of progress like this: In 1956, the new transatlantic telephone cable, using the latest compression technology, could handle 50 voice circuits. Now, optical fibers carry 85,000, an improvement of 170,000 percent. We cannot imagine it, and yet at the same time, we soon manage to take it for granted.

We shall derive philosophical profit from demanding that our imaginations show us a few consequences of this revolution. Consider for starters this description in an article called "Mass Customization" sent out by a firm of "headhunters." (Italics are mine, and my comments appear in brackets).

For an expanding number of industries, the world is *an increasingly turbulent place.* [Accelerating speed plus increasing complexity mediated through fundamentally unchanging human nature = inevitable drag causing "turbulence."]

—Customer *wants and needs* are changing and fragmenting [They acknowledge the anthropological base and distinguish wants and needs, while emphasizing HTX individualism. "The primary stakeholder is the *individual customer*-not an amorphous market out there somewhere, whose unique needs and desires have to be understood, met, and exceeded at every opportunity." But they see human nature not so much as a constant but as a puzzling source of "*changing* needs and wants."]

—Product life cycles are getting shorter [in many instances you had better recover your R&D and set up expenses in eight months or you have worked in vain! Hence speedy data gathering. A company which does not change its computers every two years (*sic*) may find itself at a hopeless disadvantage.]

—Keeping up with technological change is becoming more difficult. [You had better not be encumbered by a molasses bureaucracy!][50]

—Basic quality and service level requirements keep going up, and

50. The first board exams in cardioelectrophysiology (the electricity of the heart, subspecialty of cardiology—400,000 die yearly in the U.S. from arrhythmias) were offered six years ago, on the basis of an electrophysiology textbook with 115 chapters, not bad for a field created twenty years ago, made possible by advances in signal enhancement, fiber optics, etc. . . . hard to keep up, you say?

competition gets tougher every year. [A given level of quality becomes cheaper as automated machines become more accurate; and while the initial cost of setting up some operations has ballooned (in real dollars an advanced microchip line has become probably ninety times more expensive than thirty years ago, from, say, $10 to $900 million), the productivity—total cost per million instructions performed of the chips produced in a given year has contracted even more.]

—The companies which are responding successfully have discovered the power of *Mass Customization*, throwing away the old paradigm of Mass Production whose focus is efficiency through stability and control. Their world is no longer stable and cannot be controlled. . . . [The response is] by creating variety and customization through flexibility and quick responsiveness. . . . The journey must start with *knowledge* and *desire* . . . it advances through a *vision* that excites and energizes all concerned to reach the destination and through a *strategy* that tells them how to proceed, yet remains adaptable to changing circumstances. [So managers do control after all, matching the pace of change to the change in the marketplace]. Finally, the journey's success is determined by how well people *execute*.[51]

The anthropological dimension is certainly not forgotten in this description!

The author cites, among other examples, Lutron Electronics Company, which designs and manufactures over 11,000 different lighting controls, 95 percent of which have annual shipments of less than 100 units—and will modify them in most any way a customer desires.[52] And from the service industry, TWA Getaway Vacations, which purchases the components of vacation tours in bulk, then mixes and matches them to create tour packages that meet the needs of any particular customer. Both examples suggest the central role of cheap information: Lutron probably has most of its customers on-line, they can send their requirements directly to Lutron's computer; Lutron engineers can then on the computer mix and match preset instructions for automated machines, order up inventory, and proceed to instruct the machines, following a central coordinated schedule worked out with the help of a computer scheduling program. Quasi-automatically, and very quickly, the customer will be informed of the shipping date, which

51. B. Joseph Pine II, "Mass Customization" in *Point of View,* summer 1994. The article originally appeared in *Harvard Business Review.*

52. That sounds more impressive than it is. If a pizzeria offers a choice of 3 kinds of crust and 18 toppings, you may choose then over 1,200,000 possible combinations, which *Domino's* has no trouble assembling one by one and delivering in thirty minutes!

will be in a few days at most. Note that the fairly recent technology of computer-controlled machines of great agility is as essential to this operation as is the information system. And so is the ability (carefully acquired—through education—mind-sets) of the engineers, using new object-oriented software to work out program modules for various basic steps that go into many of the 11,000 different end designs and ways of combining them and fine tooling them into final designs. That aspect of HTX technology should not be underemphasized: the ability to accumulate, store, and retrieve an intellectual capital in the form of design components worked out bit by bit. That "fast future"—order in 10 A.M. Tuesday, product shipped 3:30 P.M. Thursday—depends on *a rich (recent) past,* stored (inexpensively) so as to be readily drawn upon.

Note in the TWA Getaway example how lack of information on one end leads the hotels, the airlines, the car rental people to be willing to sell en bloc and cheaply. (That also allows them to engage personnel and equipment for the entire season—the human factor requires some stability.) On the other hand, the retailer must keep costs in line as he works, customer by customer, to make up individual packages that fit their "wants and needs." TWA has to advertise and otherwise bring in thousands of individual customers and then devise ways of customizing those packages and sending them away happy without TWA Getaway personnel burning up too much time doing it. (Retailers who are not good at keeping track of their costs soon find they have many satisfied customers and an unhappy bank!)

For all the enthusiasm for "customization," considerable "mass production" obviously still remains in the system, at the semifinished product level at least: the hotel and airline plan for the season, Lutron has semifinished products in the wire, relays, prefabricated switches, and insulating material. *It is typical of the HTX mentality to get so enamored of the new as to neglect to notice the continued validity of the old "tried and true."*

Reflect for a moment on McDonald's as you reread the "Mass Customization" article's HTX *"Individualist Manifesto":*

> The primary stakeholder is the *individual customer*—not an amorphous market out there somewhere, whose unique needs and desires have to be understood, met, and exceeded at every opportunity.

This wildly successful mass producer of fast food has been slower than molasses when it comes to introducing innovations into its menus. Using economy of scale principles, it can push its suppliers to the wall with, for instance, five year, three zillion bun contracts, which allow the baker to order a state-of-the-art factory, cutting his costs to the bone. Within this rigid framework, a minimum of occasional variety is introduced—a new zingy sauce and a Mexican name for the "OOLAY Burger Special," with the same old fries and same old Coke; only after careful experimentation is there the great breakthrough to fast pizzas in Canada.[53] A competing Canadian chain, at slightly higher labor costs, offers "customization": You tell the hamburger constructor what (within the rigid limits of eight condiments) you want on your burger—and it is broiled, coming to you instantly without drying out under an infrared lamp.

Many people seem consoled by the knowledge that at a McDonald's there will be no surprises: the food will not sink below a certain (rather low) standard (but also, unfortunately, it will not be slightly more delicious on days when the chef is inspired). Amidst all the HTX change people like the familiar: See how they come to be comfortable with the characters and plot lines of a long-lasting sitcom, they are like old friends, until, one day, inexplicably, viewers tire of them. But then people tire of old friends, too. They like the familiar, but they like something new, too. So McDonald's almost always has one "for a short time only, hurry! hurry!" special, made from mostly the familiar ingredients.

Speed and information are important ingredients at McDonald's, too—the customer does not want to wait, the company wants him pushed through the facility as fast as possible, processes are automated to allow minimal handling and hesitation, with beepers sounding as menacing as windshear warnings in a cockpit (the whole process being a model of nineteenth-century mechanized "mass production," which, *pace* "Mass Customization," still earns good profits in fast foods!) and the customer is pressed by the people behind to make a quick selection from the very limited menu. Dining is not leisure time; nothing in the design of the airy, clean restaurant with uncomfortable seats and tiny crammed tables and employees buzzing about cleaning up trash entices one to linger. Your rent per five dollar meal is for seven minutes tenure (however no employee will ever even hint that you should leave) while Constantinos in Patras nurses his 80 cent Turkish coffee for two hours and seven minutes.[54] And both his café and your

53. Irony: Muscovites throng to experience the exotic—the banality of one is the exotic of the other.

54. Taco Bell has discovered something even the great Mr. Kroc did not think of: It offers free refills of soft drinks. A former Taco Bell employee shares the secret discovery: 1) you are paying more for the cup than the actual stuff inside it, and 2) you sit under fluorescent lights longer and buy more food.

McDonald's occupy prime real estate! (Good thing the profit margin on the retsina and the ouzo is much higher than on Cokes and fries!) Oh well, the anthropological and sociological base of the two institutions is rather different. Meantime, all supplies and sales are automatically tracked on the computer, McDonald's knows input and output at every restaurant every day. Angelos, the café owner in Patras, is happy if there just is good money left over at the end of the month!

Not only are cultures different, but the same man lives in different worlds: Jim, who is always after the latest innovation in PCs, does not want menus to change too often at his favorite restaurant. But he does want a little change. It never occurs to Constantinos that his café could change.

Thus far we have poked around at examples from just three of the many HTX sectors: manufacturing and its close cousin, fast food, and travel service. Volumes have been written on the role of "IP" in these industries. And just imagine what can be learned by reflecting on, to take a military example, the challenge of "interfacing" the information gathered by the long-range radar controlling an air battle and the on-board search and lock-on radars with the aircraft's performance capabilities and the pilot's physical limits and ability to take in and judge quickly all the information being thrown at him. Here surviving the battle is a most basic "desire," and speed and maneuverability, as well as effective firepower, are critical. Cost, while always a factor, cannot be allowed to dictate the production of an aircraft likely to lose against the opponent, just as in the manufacture of a chip, plant cost *must* include near-perfect quality. As in every human situation, some information is vital, no cost is spared to see that it arrives in timely fashion: the pilot had better know when hostile aircraft are getting into potential "lock-on" position, and he has to know when dwindling fuel dictates disengaging.

Or consider something closer to a "pure information" situation, in investing, for instance. So advantageous is it to have sound information about a company, including what is happening in its market, that fortunes can be made by obtaining possession of inside information other potential investors do not have. Here is an instance among many where legal restrictions are put on the use of true information, for moral reasons. Postman is right: institutions do control flow of information, often with good moral justification.

In summary: *context*—the larger reality—dictates what is going to count as *relevant* data; on one's grasp of the larger context depends the quality of the judgments of integration of data into the larger structures within which it gets its fuller meaning, and the worth of that grasp of the larger context is determined by the *truth* of one's judgments about it. "Truth" here means the *genuineness,* the good factual base of those large integrative judgments, which brings us full circle back to the reality of the basic data.

Reality! Information, we have just recalled, is supposed to connect us to reality. But what is the HTX doing to our sense of reality? The perennial question "What is *real*?" takes on new dimensions.

Part Two
How to Survive

Chapter 5

▼

Don't Lose Contact with Reality ... but First, What Is Real?
Into the Dreams and Reality of Social Engineering

> I refuse to be intimidated by reality anymore.
> After all, what is reality anyway? Nothin' but a collective
> hunch. My space chums think reality once was a primitive
> method of crowd control that got out of hand.
> In my view, it's absurdity dressed up in a three piece busi-
> ness suit.
> I made some studies, and reality is the leading cause of
> stress among those in touch with it. I can take it in small
> doses, but as a lifestyle I find it too confusing.
>
> —Trudy the Bag Lady in Lily Tomlin's one-woman show,
> *The Search for Signs of Intelligent Life in the Universe*

"The Truth, the Whole Truth"? or Rather Only What I Want to Know?

It is important for health, safety, and, indeed, long-range happiness not to lose contact with reality, at least not for long. (Short-range plea-sure is all too often associated with temporary deliberate loss of contact. The film industry depends on this, substituting a pseudoreality to dis-tract from a less satisfying real world. Alcoholics Anonymous can tell you something about it, too. Man's mythmaking and magic, since their earliest traces, were about both managing and escaping hard realities).

Not losing contact with reality requires continuing to integrate in-formation *critically,* not using a pat "ideology" as basis for a "triage" of what you are willing to take seriously and what you choose to dismiss. No, contact with reality requires integration of data into the large con-text of one's total reaching out to every kind of "other" that is "out there," with no evidence being shunned.

From that declaration, you can see that keeping contact with reality is a rather severe ideal, because it demands subordinating all our de-sires to a prior and fundamental will to get at "the truth, the whole

117

truth, and nothing but the truth." The psychological reality is that we all cop out frequently into ideological "fixes." But, both ontologically and, I shall argue, morally, we are wrong every single time we thus turn our backs of reality.

The challenge starts at the most primitive level: Even to want to acknowledge there is an autonomous Other "out there" demands a certain generosity, a conquering of narrowing "self-centeredness" *to be interested in what is beyond my control*—worse: that Other who is *always making demands on me!* True appreciation of the Other requires precisely not wanting to reduce him or it to a parody that will fit neatly into a set of pat answers maintained in the interest of false security— an illusion of control—and comfort (that is what I am symbolizing by the abused term *ideology.*)

The trouble is that beyond that simple truth, that we should be open to all reality, accord as to *what is real* is not always too easily arrived at. And the developments from modernity to HTX phase two have made matters much more confusing. Not everything that is out there, nor all the data presenting it, is real in the same sense, some has been deliberately arranged to mislead; and certainly not all is equally interesting, especially as agendas—what people really want—differ. All day long we select the data to which we are going to pay attention, in function of our projects. So much that we could learn is largely excluded from consideration at the start. Apparently our addicted friend considers that the reality of the present urge for a cigarette is a lot "realer" than the long-range threat to her health, which she acknowledges, when pressed, also enjoys a certain truth. Whatever that truth is, it is obviously not real enough NOW to cause her to forego a single weed. The essential role in judging probabilities of time and of freedom—of what she wants, now, versus what she wants over the long term—at work in this deployment of interpretative being is clear: She does not deny that the cigs are having certain effects on her body now that will *probably* cause bad things to happen in the future. Without giving the matter much thought—she does not want to risk that—she weights the certainty of the present temporary calming of the urge favorably against the scarier probabilities, and chooses to console herself by embracing optimistic possibilities for escaping her fate. For her, the future belongs among "virtual realities." *It is not "really real" like the present wreath of reassuring smoke.*

The huge increase in our abilities to create and manipulate data that has occurred as modernity became HTX reveals this about our freedom: Much of our freedom consists in being able to decide what to regard and how much weight to give each of the kinds of reality (e.g., various aspects of the present versus probable, or even certain, future). We enjoy a perverse freedom of denial even of unpleasant pressing realities, and it is always easy to shove future possibilities into the dustbin of the barely conscious, and inconvenient realities encountered in the past can, with various degrees of success, be forgotten.[1] To be sure, the cruel reality of the insistent "out there," both of social *Mitsein* and of nature, crashing in on the present and overwhelming everything, has the final word: "So you see, you ninny! You are dying of lung cancer just as everyone told you would!" And often the suppressed past turns out to be still "in there" somehow.

Dealing prudently with the temporality of reality is a key to wisdom: not only is the past—both the psychic and the cosmic—still real, insofar as it is effective now as source of mind-sets and of concrete possibility within and without, so is the future, as pro-ject and as the already determined to become in the inorganic and organic processes of nature, including what in nature has been stamped by human culture, like slowing decaying monuments. The in-tentional (literally "tending toward") opening and persistence on a course of action, mightily affects the present and carries one forward in a certain decided-upon direction. (A student's decision to pursue medicine, held out always in front of him, affects massively his everyday schedule). But also as he puffs his way through medical school, certain chemical effects which the cigarettes have already inflicted on his body have set in motion a deterioration that, if not reversed, will lead to results he either does not know or simply refuses to think about. If our powers of denial permit simply refusing to think about clear, present, unpleasant evidence, imagine how easy it is not to do the hard work of thinking out ahead the very likely consequences of present decisions that we prefer not to think about.

Conceptualization: One Remove from Reality, Hence Not "the Whole Truth"

From his first conceptualizations man was already partly one remove from immediate reality: he would abstract a general notion from the concrete data offered through his senses. This data could be both external (data from things and settings "out there") and internal (data about his own inner states); and these can combine. I say "partly" re-

---

1. Addiction experts recognize the central role of denial in every addiction. The line separating innocent acts of escapism from neurotic denial is a kind of self-enslavement in which part of one will not allow the freedom-center to face some aspect of reality and cope with it—it is thin and fuzzy.

moved from the primordial reality, because, although he can get lost in his concepts, the feel of concrete inner states and the bombardment of the external senses by alien realities continue unabated.

With the invention of first language and then writing, man became potentially twice removed from reality: Now there is the primordial given, the abstraction, and the symbol used to express both. Now man could listen to or read a report of what someone else had experienced and abstracted. Finally in the HTX it is still worse. Now we can get several more removes from immediate reality. Consider this example:

The other night on television an elderly Toronto lady was displaying a ball of frozen urine, which she keeps in her deep freeze, to prove that the new hole punched through her roof into her bedroom did result from this unsavory meteorite dropped from a jet. It is a fact, not a fantasy, that you too may receive into your bedroom an unwanted visitor from the world of aviation. But your new inability to sleep from worrying obsessively about this would be "crazy," although your sleeplessness and the diagnosis of it—"a neurotic concern out of all proportion to the risk"—would also be a true fact, a reality.

See the clearly distinguishable kinds of "reality" visible in this simple example. At the most fundamental level we have the factual evidence, a view of the Thing, the pee-ball. In the present example we access it, one removed, through the TV picture, difficult but not impossible to fake, but in the context we would likely judge the TV people have no incentive to make up such a story. But just now as I write I do not have access to the Thing, less so the event, and not even to the picture, but only to a memory of seeing the lady on the TV holding up the Thing, encased in a transparent plastic bag, and testifying as to what happened, while the camera zoomed to the rather impressive hole in the bedroom ceiling. And all *you* have as a source of information about the reality of this alleged event are 1) this printed version of my words invoking 2) the event of my seeing on TV 3) this account of 4) the alleged foundational event, the crashing arrival of what we are being told is the airline's little late-evening gift. You are five times removed from the underlying reality, the actual past event, and lies could have crept in at each level. (Come to think of it, until now I unwittingly just accepted the explanation, no evidence was offered for the pee-ball's aeronautic origin, it seems to be a case of "Well, what else could it be?" As I can think of no other possible explanation of this urinary meteorite, I just go along with that supposition.)

The possibilities for deliberate distortion and manipulation are enormous when "truths" are mediated through so many levels of reality.

Most Reality Comes to Us by Faith

Most (I would guess at least 90 percent) of what we know is handed us, witnessed to by agents of one or another tradition (*tradere* = to hand on). Faith is essential to the reception of truth witnessed by others. How many scientific experiments have you done to verify all that you know from the witness of the scientists? And remember that the scientists approach the realities they are dealing with selectively, with their own biases. We stand several removes from most of what we believe to be facts. Assuming your judgment is correct, your forming the knowledge that "that poor Russian immigrant lady got a pee-ball through her ceiling from a plane washroom" puts you in (indirect and complex) contact, not only with the reality of the concrete event that happened, but also with the possibility for forming now the abstract general judgment, "So, frozen urine chunks can fall from planes," which then authorizes you to fear quite rationally the possibility of such projectiles arriving through your roof. That general ("abstract") truth is formed through insight into *the meaningful structure* of such an event, leaving behind details particular to that one happening. There was nothing reported that suggests it was unique, although all the media attention suggests it is rare, and although you do not know just what circumstances cause this event, there is thus far no reason to judge that it could only happen once. In your concept of this kind of event, you leave the question of causes vague. For instance, what type of aircraft dropped the unwanted visitor? Is this process limited to that type, and under what circumstances?

Truths taken on faith are often "objectively" true. Most truths are not perfectly "verifiable," and some verification may turn out to be illusory. The fact that my present written text is a medium for your access to the event of the TV report, which, in turn, mediates the lady's report complete with pictures of ball and hole, requiring on your part three levels of acts of faith, takes nothing away from the claim of "concrete objective reality": This precise determined event, it is claimed, really happened. *How you know,* and how *sure* you can be, are quite different issues from *what is claimed*—the claim is concrete, clear, and, in this case, about basic "empirically verifiable reality." No "virtual reality" here at the base, however remote from the factual event you may be.

Contrast this process of grasping the truth about a real concrete event at several removes from the event with the "virtual reality" that simulator designers make as realistic as possible. That "virtual reality" is concrete, immediate, "empirically verifiable," (the airspeed indicator and altimeter tell of an overly rapid rate of descent; ease back on the throttle, and, there! the indicator says you have lost twenty knots) but, under the circumstances, experienceable as "so like the real thing, you really get swept up in the action, and start to sweat." It is all made as "immediate as possible, so real you start to believe it." And yet no pilot would mistake, as a sick person does in an hallucination, the "wonderfully realistic" virtual reality for a real "Mayday!"

It will help our efforts—necessary for survival—to keep clear about different kinds of reality to see that *immediacy or remoteness has nothing to do with the truth of what is experienced or claimed.* In the imaginary world and the conceptual world, the one able often to be more concrete, the other abstract, the subject is one or more levels removed from direct contact with sensed reality. In virtual reality simulations, on the other hand, one is presented with a flow of sensations programmed to produce a response *as though* "the real thing" were immediately impinging on him.[2] It is true that a pilot experiences (hence with immediacy and concretely) the (artificially imaged) runway looming up at him too fast, it is false that the runway is real and that a plane is about to enjoy an unfortunate meeting with the asphalt, he is not about to die. It is true that a psychotic patient "sees" and believes in the reality of the horrible black figure ordering him to kill. It is false that there really is such a personage in publicly accessible cosmic *Mitsein* time-space. It is true that exploding M_3 money supply can cause inflation, although the conceptual abstract relationship just expressed is unexperienceable. But some of the concrete fallout from the extreme inflation, like people losing their jobs, will be palpable.

The ultimate in producing a virtual reality effect would be to place electrodes in the brain, producing a compelling hallucination.

Learning to Survive in Virtual Reality

I used "virtual reality" in the title of this book as a way of posing a central kind of HTX problem. For, as we have seen, the HTX complicates in many new ways our relations with reality. The time has come to distinguish several kinds of "virtual reality":

2. When I see my virtual image in a mirror, through the medium of the light reflecting off my figure and then transmitted through the glass, I see me, in the Gestalt of an image reversed from what you see. My cat simply cannot believe that image is her; nor does she think it is another cat; in fact she reacts to it not at all. I remain fully aware, as perhaps an infant is not, that I am seeing my (reversed) image and not directly me. In an hallucination, something happens in the brain to present a kind of virtual reality concrete presence that appears, somehow, out there and is terrifyingly realistic.

1. "Engineered virtual reality effects" are *deliberately arranged sensible effects meant to produce an immediate representation*. This includes not only simulations intended to produce pseudo-experiences either to train or to entertain, but also computer simulations that permit manipulation and experimentation.[3] I shall term this form of virtual reality *simulation*.

In a realistic horror movie, the sights and sounds are indeed programmed to produce a direct effect of "reality," and the filmmakers would love to be able to afford to rock you in your seats and spray odors into the theater to complete the effect. Movies and TV plays, and for that matter Walt Disney's Magic Kingdom are virtual reality phenomena. A realistic photograph and a clip on a TV newscast are truncated virtual realities, meant to provoke credence: "What do you mean it did not happen? I saw it on the TV news with my own eyes!" When "Rock music" pounds away at your most primitive instincts, the producers are engaging in something of the same kind of mobilization of your outer and inner senses to summon up a kind of imaginary reality. The spiderweb-looking first 3-D design of an airplane fuselage allows us to see, and to work with, configurations that do not exist, and to change them with the click of a mouse.[4]

This raises a question: Do we, in healthy states, always have to agree "to play the game" in order to enter a virtual reality world? If so, would it not be the case, then, that involuntary "virtual reality" effects are symptoms of mental illness: delusions, illusions, and most strongly, hallucinations?

The term *virtual reality* is sometimes used for something like the Internet "community," which is spooky: It is what it is at any one moment, but changes incessantly: At this moment there may be 123,145 people on-line—each of them an incarnate human being existing in cosmic time-space; fifteen minutes from now there may be 83,560, of whom only 34,214 were on-line fifteen minutes ago, but then no one

---

3. Either with things that do not yet (or may never) exist, as in design of futuristic tools, or kinds of manipulation of things and situations which do exist but to which we cannot have the kind of real access the computer programmers can imagine, and of which they make possible a computer simulation. Nuclear engineers work hard never to have a real hands-on experience of a core meltdown, but in computer simulations they can cause a virtual meltdown and play with its ramifications.

4. The differences between these distinct kinds of manufactured experiences are more of degree than of kind, because all these tricks seek to fire mechanisms within us that summon up images and provide the emotional fleshing out that normally accompanies our encounter with real persons and things and thereby to substitute an effect for the actual presence of the real person or thing that normally produces something of that effect.

really knows. I shall term such a *floating community* just that, floating reality. The HTX is not simply a floating reality, because enduring structural features, well anchored in human inculturation and institutional arrangements and structures, give it backbone. Indeed, no reality can be entirely floating. Even the superdynamic Internet requires protocols and a communications trunk that remains fairly stable, even the shortest burst of radiation has an amplitude, frequency, and duration, be it a fraction of a nanosecond.

2. One hears talk of the "virtual office," or the "virtual university." To the extent electronics allow one to operate from a motel room as though he were situated in a conventional, unmovable office, there is good reason to speak here too of virtual reality. I shall call this form of virtual reality *electronic presencing*. As in a telephone conversation, a new kind of space is produced, based on annihilation of some of the effects of physical separation.

The verbal interaction on the telephone with loved ones provides an audible presence that the imagination fleshes out with an inner visual image. When we have never met the person, we involuntarily supply an image to go with the voice and style of speaking. This mixture of highly reduced reality and imagination constitutes a less intense level of involvement of the whole person than simulation, but a more genuine interaction—that really is my daughter I am talking to on the phone. (When I was an Air Force lieutenant in Air Defense we used to while away the wee hours of the morning talking from Watertown, New York, to the RCAF ladies of St. Hubert, Québec, with their lovely French accents. When the invitation came to drive up and meet them, we decided not, fearing the reality would never measure up to the fantasized version.)

3. If we adopt the strict usage just suggested, then a wonderfully evocative realistic novel should not be said to present us with virtual reality, because it passes, without the aid of electronic devices, and by way of language only, to provoke the imagination to provide the necessary *inner* sensory images, but never with the realism of a delusion or hallucination that something is presenting itself to us "out there." We are clear that the figures and events in the novel are imaginary. To be sure, once the reader allows them to surge up, their inner presence can produce a tingle of fear or a sexual arousal. The great novelist can provide us with a depth and richness of presence of his characters never attained in a film, which must work through sensations and spoken words, evoking situations but cannot, except in soliloquies, provide the inner life. In neither film nor novel do we mistake the characters for real persons present before us. They inhabit a kind of imaginative "virtual reality" space of

their own. I shall call this form of virtual reality *verbally aroused imaginative space*. The intervention of language introduces something quite different from the immediate effects essential to creating virtual reality of the simulation kind. When the mystery story you are reading uses verbal tricks of every kind to get your hair to stand on end, that is not an immediate impinging on your fleshly emotions, but a mediated one, through language. By contrast, I look directly at the spidery three-dimensional image of the airplane fuselage on the computer, or I am immediately assaulted by the screaming voice of an angry friend on the telephone. A concept, or a mathematical or verbal symbol stipulated to stand for a concept, substitutes for, and often refers to, real things, but does not necessarily seek to invoke an aura of their presence by provoking an emotion. Analyses of real things, persons, or situations into parts, and the abstraction of parts and the generalization of arrangements so that they become held as "typical" or as "general possibility," are removed from reality, refer to it, but do not attempt to invoke the sense of a concrete presence.[5] But like all cognitions, including virtual reality experiences, as described in "simulation" and "electronic presencing," they require that we refer back to experience of the fuller context, ultimately to some part of the real world if we wish to judge their *truth*.

Poetry forms a special case. Unlike the analyst or the scientist, or everyday man simply informing someone about a state of affairs, the poet plays with the concrete evocative capacity of the words, wedding rhythms and images, to provoke feeling approaching music. This is the closest language can come by itself to having something slightly like a "virtual reality" effect.

Distraction through virtual reality entertainment is morally not much different from the more classical escape into novels, or for that matter, tourism, where we wander with no real cares beyond getting unpacked and not missing the tour bus, through strange landscapes filled with foreign people with whom we enjoy only carefully managed contacts. There are many ways to use parts of reality to detach ourselves distractedly from the whole of it.

Hierarchies of Abstractions Confuse Our Sense of Reality

Finding the larger reality that enfolds a simulation, an electronic presencing, a film or a novel, or for that matter, hierarchies of abstractions

---

5. They can, and do, of course, often provoke action, as when a reader fired up by the *Communist Manifesto* joins the party and begins working to bring about the revolution.

built on abstractions is surprisingly easy for a mentally healthy person. Still, utilitarian instruments of abstraction, like money, can be dangerous for our sense of reality. Think what happens in hyperinflation. Or, more remote from reality yet, consider "put options" on the S&P 500. What is the "reality" here, my profit or loss when the option expires or I cover the put? What relation to a larger reality does my action have? What (tiny) effect on the market? I don't know, and I don't care! I just takes my money and goes home! This much though is true: Through the intermediation of a broker operating in a derivatives market, I am betting against someone else, whom I shall never know, that within a given time the S&P index of 500 stocks will go down more than I paid for the option. This relationship is a "reality," too, but to me my loss when I finally cover or let the option expire is much more real: it is money I shall not have to spend on something else.

It is astonishing how far removed is this "instrument" from the underlying realities: 1) the events happening to 500 companies 2) as they get reported and 3) become the base of millions of decisions whether to buy, hold, or sell those companies' stocks; 4) the prices of the chosen 500 stocks being averaged as the "S&P 500 average," 5) upon the rise of which, over a given time, I am betting against someone who believes my hope is misplaced. From the minor little event: this guy buys a PC from IBM, through adding up sales, factoring revenues against expenses, announcing net, to decisions to buy or sell IBM stock, to inclusion in the S&P 500 average, to effect on the put option—that is *six levels of abstraction removed* from the level of manufacture and sales! On each level of abstraction one is involved with phenomena that connect causally with thousands, if not millions of other human actions, to form vast aggregations, the *averaged* senses of which have been abstracted and which provide some meaning to my concrete action, the purchase and covering of my put.

To see how far we have come in inhabiting a virtual reality world, contrast that HTX transaction with a simple face-to-face barter: I have these bags of grain. You have that sheep. Do you want to trade? Our simple exchange affects, even indirectly, a very small circle of people and in a narrow range of effects, most of which are easily traceable. The movement of a vast market affects millions, maybe billions of people, if only slightly.

And yet for all this complexity, remoteness, and widespread effects, the options trader has no great difficulty "keeping in contact with reality" throughout the entire options extravaganza—all that I have recounted is perfectly "intelligible"; in some way "he knows where he stands," and "what is going on," there is a clear "bottom line." No fantasy need necessarily be involved in all these levels of abstraction, but it can creep in at any point!

Already just putting almost all of our commercial transactions through the abstraction of money favors the development of a *voluntaristic* mentality, which has reached a high pitch in the HTX. When I put a price on an item in a barter—I value three of my potatoes as worth your watermelon, and you agree—the concrete presence of the real things somewhat disguises the fact that in setting a price I am *valuing,* which means I decide what I am willing to consider these things "to be worth." When, however, the transaction goes through the neutral abstraction of a monetary figure, it somehow dramatizes that these things are worth what we (perhaps perfectly arbitrarily) *decide* they shall be worth. We are far away from any consideration of their relative importance on "the ladder of being," and the truth that the most precious beings have no price—it is not moral to sell your son as a slave, no matter how good the offering price! How foreign has become the notion that things are good intrinsically, by virtue of their being! Even in an arena as subject to wild swings of taste as art, an unrecognized masterpiece can be acknowledged to be a thing of great beauty even when it has been neglected down in a coal cellar. Still today, money conscious as we have become, most people would hesitate to believe that the true worth of a person is measured by his salary. Mother Teresa got no salary, and her estate consisted of two saris and a pair of sandals. Nor does her "value" reside in what the public at any one moment may think. "Judge not!"

Managing Information: A Personal Introduction

Here is the good news: Fantasy need not creep into our view of things just because we are obliged to pass through levels of abstraction. The bad news is that the HTX settings in which we have to operate are so complex, there is often more involved than just getting back down to reality through levels of abstraction. One is obliged to discover in a given very complex situation what are the *relevant concrete factors,* how to discern their true makeup, how to judge in what direction certain processes are most likely to evolve, and then relate these givens and *balance them* to build up an adequate picture of a real objective unfolding situation.

Wisdom requires not only a balancing act, as we saw earlier, but also selectivity.

It helps to go anecdotal to see what all is involved. Take a real-life example of such a struggle to balance: Recently a friend offered to sell me his 8.8 percent interest in a small HT manufacturing company, of which I also own a like share. A well-focused, concrete question: Do I, yes or no, buy his part for the offered price of $18,000? Behind it is the question of assessing the risk and my "comfort level," not based on irrational whim but reflecting an assessment (inevitably rather vague) of how reasonable it is to place a certain number of eggs in a given basket.

When I thought about expanding my participation in the company, I was nervous about my existing exposure. I knew that the company was involved in some promising negotiations. But here is where a sense of balance comes in: I judged that I have already too much of my resources riding on this little, fragile company's success, and so, even though the opportunity looks pretty good, because there are reasons why things can still go sour, I will not risk more on this one venture.

This sense of balance in the context of my total investment situation rests, as I admitted, on pretty vague evidence, but it is neither per se abstract, nor imaginary, in the sense of simply made up. I am looking over my entire portfolio, a concrete assembly of investments, assessing the degree of risk of each, and then relating the various risk levels to one another, and all this to present and future assured income, and present and future estimated expenses. I am weaving a tissue out of many judgments (my house is now evaluated at X thousands, and so on) in which many processes have to be related and extrapolated. Concepts are of course at work in all this reasoning: for instance, as regards my oil stock, how long will the glut last—and will my oil company last longer? A conceptualized general equation, which could be stated in mathematical symbols might be in order here. And in extrapolating there is inevitably an element of creative imagination. But despite the role of abstraction and future anticipating imagination, what I am judging ultimately is this and that concrete investment, judged, to be sure, against a rather abstract background, as a context for deciding on the present concrete opportunity to buy 8.8 percent of this particular company for this exact price. My decision is ultimately concrete, too: No, I put in no more money here. (Too bad! Four years later it would be worth $156,000!)

Managerial BEING Considered in Its Own Right

This was a good example of the typical "managerial" thinking demanded of us all by the HTX, with its balancing of many concrete factors and its mediation through many levels of abstraction before returning to concrete decision. What makes matters worse is the fact that each of us is interconnected with so many organizations, each trying to ride herd on many dimly perceived processes. Our efforts to go on with all these activities and to balance them, not only in our picture of them, but in our deployment of time and energy in dealing with them in the midst of fast-moving concrete situations, leads inevitably to that great enemy of wisdom: *DISTRACTION*. No wonder we are tempted to flee into the unreality of entertainment distraction.

So we spend endless time in meetings, we allow our most earnest conversations to be interrupted by phone calls, we seem to have little time for cultivating real friendships; burned out, we fall into some vapid form of tourism for recreation. Here is one reason why so many in authority have no vision: they allow little time *to struggle to see* in the midst of all these complexities.[6]

Being forced, for survival, to manage, hence to adopt an analytic, instrumental, and finally voluntaristic attitude, as opposed to a more receptive, appreciative, grateful, *be* and let *be* attitude, we risk *becoming* what we *do,* becoming mere managers, in every aspect of our lives. By that very fact we then risk destroying what we manage—ourselves and our "associates"—because we never "manage to be present" to the deeper, broader, more elusive realities upon which any ultimate mature human success depends.

This struggle to manage *can* be maturing. But for many it only produces an acute sense of being caught, a feeling of poverty when surrounded by riches unimaginable to the billions of peasants in the world today.

We have only ourselves to blame for much of this distortion. Truly, this efficiency-oriented, managerial HTX being has delivered to us so much possibility, it has made life so easy for us, that minority, the citizens of the most HTXed countries, we forget that we are freed up by our good communications, our ready-to-eat food supply, our good

---

6. When Patrick Haggerty was running the rapidly growing Texas Instruments, Inc., he would suddenly disappear from view for two weeks at a time."Oh! Oh! Brace yourselves! Haggerty is thinking!" the managers would say. He would return, convoke all the vice presidents for a Saturday morning session, and announce horizon-opening initiatives. (Personal account from Bryan Smith, a retired vice president).

health, our long life expectancy, so that in a certain sense we have more *disposable* time than any people has ever had. If managers do not "find time" to develop a vision, it is probably a poverty in their own character that is to blame. (Remember Jacques Vidal of the *Shell française.*) They could "make" time, and they can learn to use wisely the libraries of information at their fingertips.

"Managing" versus Commitment: What Is "Dehumanizing"?

So while the managerial being of the epoch is proving in certain important respects richly enhancing of human possibility, in others it is damaging of the human reality out of which any society is built—it is "dehumanizing."

Of course the limits peculiar to any social form in every epoch are (by definition) hampering of the fullest possible human development, each poses special temptations to distortion of character, just as each social form offers some strengths in developing maturity. The idyllic village life (still surviving at the margins) was laced with petty tyrannies and hardly encouraged great creativity—look at the snail's pace of human innovation over early millennia. But stable it was, things moved at a "human pace," most of the time (or was it closer to an animal pace?); that perennial rhythm characterized most of human existence for the first few hundreds of thousands of years and guaranteed what we perceive today to be general poverty and a short average lifespan. To be sure, the form of "dehumanization" that allows us to avoid all human contact in any depth was, and remains, impossible in peasant society. But they had other ways of hampering human growth.

It is yielding to the "spoiled child temptation" to dream of perfect societies that never existed and to dwell on the negative in our own urbanized HTX situation. Consider that modern society delivers a vaster conceptual range of possibilities and resources to make the pursuit of those possibilities feasible. Our intellectual landscape has been expanded to embrace a cosmos at least ten billion light years in extent and age.[7] Our knowledge of history has been extended and refined. And yet, 'tis true, the *dangerously hectic* life in the most advanced societies is becoming so extreme, the twenty-year-old student who said to

7. On the implications of this vast leap in our grasp of ultimate dimensions of reality, see *Being and Truth,* chap. 10, "The Ultimate Structures and the Overcoming of Ideology." Incidentally, the precise age and extent of the cosmos is not known, unless you consider, say, 14 billion with a margin of error of plus or minus 5 accurate!

me recently, "People in this town are so stressed out, I mean its scary!" is right to be concerned. Distraction may lead to implosion. All indicators of healthy social life are negative, not the least being the startling fact, which no one seems able to face, that *every advanced HTX people is dying out.*[8] Among some of the poorest and most illiterate, however, populations are still growing, producing possibly a net decline in average education in the world.[9]

What happens when the Greco-Christian-European civilizational foundation of the HTX gets so severely eroded, internally through secularization and failing numbers, externally by immigration from non-Christian lands, that Christian ethics no longer exercise much restraint, and Greco-Patristic-medieval-modern European humanistic intellectual culture is almost totally replaced by a narrow, shallow-rooted, success-oriented, calculative instrumentalism?[10] It would be Hong Kong without strong Confucian families. Just see what is happening to Russia precisely because of the moral rot of seventy years of atheist totalitarian rule.

At the base of much of this social negativity is the *general diminution of commitment.* This is a fundamental characteristic of the emerging HTX phase two. *Commitment* is freely engaging one's very existence, agreeing to a consistent and faithful investment of one's being in someone or something, which is an effective witnessing to what one believes, a "martyrdom," as one dies to self, sacrificing selfishness.[11] Commitment

---

8. A friend who has installed machinery in several factories in northwestern China tells me that it is among the intellectuals, the engineers, and the urban workers that the one child per family program is effective; among the great masses of illiterate peasants, leading life in very reduced circumstances, it is observed in the breech, indeed families of six or seven children are not rare. He discovered that in his remote province, the population, despite migration away to the south, had grown in ten years from 30 to 39 million!

9. The impact of this enormity is fudged to some degree by immigration, but that is only causing serious tensions, especially in the old European lands. The implications of this demographic earthquake are being evaded by media and intellectuals alike.

10. Marshall McLuhan, the laconic prophet of the electric age, asked me one day, as he lay on his sofa, pointing at the softly mumbling TV, "Do you want to know what I really think of that thing?" Of course I did. "Well if you want to save a single shred of Hebrew-Hellenistic-Roman-Christian humanist civilization, take an axe and smash those infernal machines." This Luddite passion in a Catholic convert did not altogether surprise me.

11. Having been suckered or manipulated into something is not, then, genuine commitment, it is a form of enslavement. Every commitment involves a sacrifice of self—the selfish self-centered self—and hence is witness to something greater, which is what the Greek word *martyrein* means.

is an essential element of response (*spondeo* = I commit), for if you offer nothing of yourself, if you do not make the act of confidence (*fides* = faith) by opening a credit to the Other, the Other who is calling—be it a thing, animal, person, or God—cannot reveal much of itself or himself. Commitment may in one case involve a very specific "performance contract"; but in another it may be an implied promise, not to do anything specific, but *to be present as needed,* "to stick by someone."

Commitment does not have to be reciprocal. A mother may never give up on one of her children, although he has turned hatefully against her and the whole family. She prays for him, and is at every moment prepared to receive him back, without recriminations, like the father receives the prodigal son in the parable. But there is no denying we desire to see our commitments reciprocated, and in many life situations, one is just being a sap not to demand some commitment in return.

Most commitments in life are implicit: from the situation the parties infer what they progressively commit to one another, as in a growing friendship, which may become very committed without either party ever actually saying, "Come hell or high water, you can count on me. If you ever need a loan . . ." And over time, and depending on circumstances, commitments can vary. It may become impossible, despite all the goodwill in the world, to maintain a level of commitment that earlier was taken for granted.

People with shallow roots, or little tradition at all, can scarcely be expected to be maturely committed over the long run or even to understand unconditional, "to death do us part" commitment to anyone. Mature commitment, which requires knowing who one is, remains fundamental to the stability that allows healthy, prosperous human existence.[12]

Now, when everything appears to be changing, it is hard to wisely choose one's commitments. Indeed the confusions brought on by such rapid change as the HTX has generated can entice people to fanatical forms of commitment to idols of all kinds in place of fundamental realities that get obscured by all the surface commotion.

If there has been a weakening of mature commitment (the divorce rate, the number of couples cohabitating, the rate of abortions, and the rates of job jumping and brutal downsizing are all indicators) on the one hand, and on the other, outbursts of ideological or just plain pathological commitment evident massively (in sects and cults and the great fanatic movements of our time—communism, fascism, Nazism), what

---

12. The SS *Obersturmfueher* was fanatically committed. That is not mature commitment. On the difference, see *Being and Truth,* chap. 7, secs. G, I, M.

does this reveal about the being of the HTX? Will the undermining of commitment favored by managerial being (extreme competition, the need "to use" people as one roars down the efficiency highway), while keeping some "on their toes" for a while, not burn out most employees in the long run?[13] (A study of "reengineered" firms showed increased productivity in the first year, followed by sagging productivity and low morale in the following years.[14]) And why does the effort to manage, through the welfare state, the damage caused by the resulting social instability not succeed better (look at the social indicators, despite some undoubted successes, especially in reducing poverty among the elderly and, in the U.S., among the strongly targeted blacks)?[15]

The present HTX atmosphere has taken on a note more sinister than just the already grave problems inherent in scale, bureaucracy, and rapidity of change. *Instrumentalization,* founded in an attitude in which basically one uses another to achieve his ends, while never absent since Cain and Abel, is now interwoven with, and feeding the peculiar being of, this HTX epoch: Not only has instrumentalization reached massive proportions, not only has it become systemic and virtually inevitable, reaching from conception (I choose whether this new life will continue) to death (I choose when to dispense with this now "useless" old person), it has come to define too much of the being of the instrumentalizers themselves.[16] Without reflecting on it, HTXers are slipping into treating *themselves* as mere instruments toward some disincarnate end of a superhuman society.

---

13. In contrast, the Sun King—Louis XIV—may have "used" people at his court, but wholesale dismissal was difficult, given that they were born into their rank! The greatest irritants he could cast into prison. But in all he had to be careful. The court was never brutally downsized.

14. *Business Week,* September 1995, exact reference lost, article on reengineering Corning Glass.

15. Indeed, not everything in this regard is as bleak as the archconservatives make out. Consider these signs of progress achieved by blacks in the U.S.: In 1970 14.2 percent of blacks aged 25 to 44 had spent time in college; by 1990, 44.2 percent; high school drop-out rate has fallen to 5 percent, about the same as for whites; black men in high-tech jobs make almost as much as whites; and male college graduates show almost the same percentage of blacks as whites in executive positions (28 to 30 percent); black college-educated women earn $28,288 versus $27,088 for white women (last census); blacks have done better than whites in stopping smoking; and although 3.1 percent of 18 to 25 year olds were crack users in 1991, in 1993, only 0.7 percent; in 1966, 41.8 percent of blacks had incomes below the poverty line, in 1993, only 33 percent (but that is still twice the rate among nonblacks). *Economist,* July 8, 1995, p. 27.

16. That "too much" is obviously a judgment based on a non-HTX faith. More justification will be offered in the next chapter for this personal judgment.

Large institutional structures drenched in managerial being, founded in partially false beliefs about what life is meant to be, form "an occasion of sin," an almost irresistible temptation to instrumentalization. As human relations become ever more mediated through distancing abstractions, instrumentalization becomes yet more massive: The huge scale of markets, which in turn dictates the scale of some enterprises, and of national governments has affected the very structure of mass public education, and has brought on multilayered instrumentalization. It thus becomes ever easier to fall mindlessly into it in making daily decisions.

Critics of capitalism are long on condemnations of all this and short on credible alternatives for organizing mass industrial society. Within capitalist circles there is increasing interest in reducing governments' roles, "devolving Federal power to states and municipalities," "flattening management structures," and "subsidiarity" (a term found in the Social Encyclicals of the Church, meaning leaving the lower structures free to do their job [by, for example, giving parents education vouchers and favoring volunteer social work organizations]). Most people now admit that socialism—central state capitalism and control—has produced even more sinister forms of such instrumentalization than free markets.

While I as manager may be interested in George only as someone who buys my product, and in Jimmy as someone who sweeps the plant, it so happens that in reality these instruments are also persons, and *the whole person is engaged* in the most compartmentalized, simplified, and superficial act. So, realistically, even as manager I had better stay aware of the nature of that instrument. When it feels abused, it flares up. *Cet animal n'est pas méchant; mais quand on l'attaque, il se défend!* warned LaFontaine . . . and such an instrument can often feel himself "attacked" without our having a clue, leaving us bewildered when the flying fur finally settles.[17] (When whole classes of people flare up in bloody revolution, those whose worlds get torn asunder overnight never seem to have seen it coming or to know what hit them.)

If scale, hence bureaucracy, hence abstraction, and rapidity of change make instrumentalization inevitable, a sense of *combat to the death* makes it rawer still. When competition is healthy, one can find a humane cooperation (I do not mean collusion) between competitors, as when Air Canada helped out one of its competitors, American Airlines, the other day by lending them a fuel tank lid from a DC-9

---

17. "That animal is not mean, but when attacked, defends himself!"

which also fit an MD80. Enlightened self-interest? Next time I need a part, you'll help me? Or just human decency? But when people are being *taught to hate,* as the class-conflict, politically correct mentality of resentment does, then you wait for the other guy to be down in order to pounce. That attitude is not conducive to health, unless your concept of good human health is based on some raw Darwinian ideology about the prosperity of the most brutal survivor.

In the example of the comradely sharing of plane parts between "mortal enemies," note all the stability there is in the relationship: We are all flyers, we have shared the same airport for forty years, we have a tradition of helping one another out, and the mechanics probably share their love for their métier over a beer. When instability reaches a certain degree, this becomes impossible. Lamenting the loss of the traditional office for the virtual one, a retired *Business Week* editor, Jack Patterson, points to the destruction of "intangible but indispensable values I discovered at work: the sense of community, the shared goals, the spirited exchange of ideas, the pride of achievement—no more standing around the water cooler, exchanging stories and maybe learning something useful."[18]

"Almost every employment relationship is contingent on the employer being able to continue in the same business with the same product and the same technology with the same quiescent competition," says Audrey Freeman, a former Labor Department economist.[19] "Now who is in that position?" she asks.

As one peruses a popularization like the *Business Week* special issue, "Rethinking Work," the inherent trends imbedded in HTX being spring out with startling clarity. They spell, negatively, more instability, less security, more anonymity, less comradeship, narrower analytic focus, more cruel instrumentalization—all dimensions we have observed accelerating since the beginning of the industrial revolution, and seemingly irreversible. But then, in fairness and balance, recall the positive benefits: high productivity produces wealth, that wealth is spreading throughout the world, it buys HTX medicine and better nutrition, permitting more human beings to live longer and more comfortably, and to enjoy ever-wider opportunity, a throwing open of horizons, a liberation from the narrowness of peasant life; more and better quality information can be quickly gathered and further disseminated from around the globe. Never has access to art, literature, and philosophy been so available to the masses.

18. *Business Week,* October 17, 1994, p. 87.
19. Ibid., 77.

While one acknowledges with the *Business Week* editor that "people can find almost as many ways to force technology to adapt to them" as technology has found ways to force them to adjust to it, the dark question remains: Will the being of the HTX produce such splits between the "haves" who can handle the necessary education, adapting and coping to the stressful situation, and the "have nots," who simply cannot hack it?[20] And the have nots may sometimes be brilliantly educated persons without the "dog eat dog" temperament necessary to survive in managerial society. Is there not enough *anomie* growing even among the haves to cause rending in the fabric of their lives, compromising at its base the whole great hierarchical, abstraction-mediated social structure of the HTX?

"The poor you will have with you always," Christ reminded His disciples. Note the context: He is putting the group CFO, one Judas Iscariot, in his place for complaining when Mary Magdalene "wasted" precious oils anointing the feet of the Son of Man instead of giving the money to the poor. Lots of lessons there. The problem then, as now, was not just with the poor but even more with the rich. Within the highly HTXed countries the marginalization of those without the education required for the interesting new jobs is of course a burning problem. It has been since the beginning of industrialization, and it is getting worse as the percentage of the workforce requiring ever-higher skills increases. When a large enough "hard core" of unassimilables builds up, and external emigration is closed to them, they emigrate internally, forming dangerous cores to the great cities, where they are *systematically* taught a rhetoric of criminality and offered on a silver platter opportunities for serious crime. This "internal proletariat" (Arnold Toynbee's term for the dangerous unassimilated group in every society instrumental in the ultimate destruction of that society, and even whole civilizations) becomes, locally at least, dangerous to life and limb.[21] When the profits from the crime that is not supposed to pay reaches proportions like the estimated $200 billion of the U.S. drug trade—greater than Japan's surplus—excellent brains can be purchased for organizing high-tech operations for a transnational Mafia of nation-threatening power.

The cities become more brutal, management styles "more aggres-

20. That U.S. employers expend only 1.4 percent of payroll on training, barely increasing at the rate of inflation since 1990, is ominous. How big a contribution to American inventiveness and productivity is being made by imported skills?

21. See Arnold Toynbee, *A Study of History,* 12 vols. (New York: Oxford University Press, 1961), vol. 1.

sive," managers more "stressed out," workers more productive but with stagnant real incomes, "structural unemployment" rates higher, families more disrupted by having both parents work, by more frequent community-destroying moves. The little nuclear family finds the strain of hacking it alone almost unbearable.

But here is a glimmer of hope in the gathering gloom: Might not the effects on the system of the increasing depression of parents (and escape into mind-altering substances and exotic relationships) and above all the psychological disturbance among the children be offset by the system's openness to promotion of those best adapted into the places that, in older forms of society, would have been inherited? If talent is more free to manifest itself, and to move around, even from country to country, will that not serve the society as a whole? Is immigration not nature's way of restoring a balance where old populations have grown tired?

Engineering the Social Safety Net

Meanwhile, we do have the (increasingly unaffordable) social welfare net, do we not?

The assertive large-scale "social engineering" visible in the erection of the welfare system is typical HTX. In contrast, family- and faith-inspired charitable works are not only one on one, but much less planned and calculated: You see a need and generously, and out of a sense of personal duty and responsibility, rally around.

In the past, by our standards, those societies of instant charity were all poor (the members of religious orders who ran many of the hospitals and homes for the aged in Catholic societies actually took—and still take—a vow of poverty). In judging how well they took care of their most dispossessed—"the sick, the widow, the orphan, the prisoner"—an HTX scholar needs to do a rough calculation of what percent of the meager GNP was devoted to helping, and he must find some good criterion of effectiveness. Mother Teresa has been criticized for showing no interest in social engineering to overcome massive social injustices. Her response is, roughly, "I applaud those who deal with such vast problems. Meanwhile my sisters and brothers and I have no plan, we just see an immediate need, roll up our sleeves and deal with it with whatever God offers us to share." The HTX endeavor to embrace in THE planning vision an entire segment of society (e.g., health or pensions) and to create *ex nihilo* grandiose solutions has something in common with the mentality that stole rivers and dried up

Aral seas. An astonishing creativity has led to the invention in the U.S. of close to three hundred federal social programs of one kind or another, and while, as I acknowledged above, one area of poverty was vastly diminished, that of the old, and another considerably improved, that of the blacks, yet another grew far worse: the number of single mothers leaped, a disaster encouraged, say the critics, by poor social engineering. And nowhere (except possibly Germany) is there satisfaction over the operation of immense health-care systems.

A huge social welfare bureaucracy with a natural tendency to keep its "clientele" dependent keeps the programs growing, just as happened with the "military-industrial complex" President Eisenhower warned about. Resultant higher government debts have meant higher interest costs, higher taxes on business and on everyone, reducing the job-producing ability of the society, contributing to high "structural" unemployment (causing victims distress and producing more work for the social welfare industry) and to stagnating real incomes. Even in the U.S., where a recent boom has lowered unemployment spectacularly and put government budgets in the black, incomes of ordinary middle and working class families have remained stagnant. All of this social engineering has been accompanied by bureaucratic meddling in private (often religion-based) social welfare institutions, such as schools, hospitals, and centers that provide care for the elderly, especially aimed at increasing state control and "secularization." The effect of all of this has been to weaken the family's incentive and ability to care for its own, especially when "its own" is thousands of miles away.

While a mentality of managing vast-scale projects of social engineering has prevailed, those most actively invested in propagating and perpetuating state social programs do not seem (and for some good reasons) overly eager to develop and apply some "bottom line" measures of "efficiency" in delivering care. Rather it is the conservative proponents of free markets who cry "Waste!" and foster studies trying to show that the private sector delivers assistance more "efficiently." Those actually delivering the care are aware that it is human time and human suffering that is most precious, and that love cannot be quantified.

Obviously, to the extent one mobilizes volunteers who donate their time, and with it undoubtedly a lot of heart, a given result may well involve the expenditure of fewer dollars. But in massive urban systems, one must be sure the basic care will be on call when needed. How do you measure "heart" and "suffering"?

You do not have to be a "social scientist" to see at once that the variables in such an immense social welfare system are almost beyond number and include many *attitudes* that are not easy to measure even approximately, *pace* all of social science. However HTX and however calculating one becomes, one is confronted in this arena inevitably— because of the level of abstraction and the dubious measures at the base—with a clash of ideologies. What is less obvious is how these ideologies all take forms that reveal the HTX being of the epoch.

It is important for understanding the HTX to see why this clash is *ontologically* inevitable, and what it tells us about surviving (and prospering) in these "virtual reality" times.

Behind Social Engineering: What Do You Want?

First, a bit of primitive anthropology: Whether I undertake to help a refugee get a start in business, or a provincial government decides to set up state-run daycare centers, the preponderant determining factor is what do I, or this government's cabinet (and behind it some segment of the existing social welfare bureaucracy), *want* to do. *Cherchez le désir derrière le pro-ject.* (To ask, "What do they really want to accomplish?" is not necessarily cynical, but given the effects of "original sin," it is always realistic to be on the alert for an element of the self-serving.)

In such a field of human relations, the driving desires can never be as clearly formulated and publicly defended as a manufacturer's goal— "I want more market share, so I am willing to suffer lower margins for this year to attain it." In the case of the manufacturer, the *measure of success* is annoyingly clear: End of the year, how much did market share increase, and what affect did lower margins versus higher sales have on profit? In contrast, who can ever be sure he knows what "the government" agents want in getting more into the daycare business?

To suggest how vague and hence ultimately ideological the motives behind all large-scale social engineering schemes are likely to be, and how, as a consequence, management decisions must be different than those taken in commerce, I shall conjure up an instance of what might be a government's desires in a particular case.

Like any group, governments have mixed motives. At the top of the list, perfectly justifiably, is their desire to get reelected. But as there are many things the government can do to jolly the reigning party's natural constituency, so there are usually other reasons for investment in one initiative rather than another. These days governments tend to throw a bit of support behind just about any initiative that will make some cohesive, pushing "special interest" group, if not happy, at least ready to cool the rhetoric of their attacks. Note the HTX way of calculating in all this. The general goodwill of the population, desiring order and tranquility through fair government, is measurable by large public opinion polls. Any more definable segment of the population, for whom some well-organized group agitates, is classified as "special interest." Their power is less in numbers than in their ability to lobby and to manipulate the media to stir the "opposition" to cause public pain for the government over a stretch of time.[22] Both sides—conservative and liberal—are forced into making macroeconomic-social judgments that are very hard to ground well; basic, little debated assumptions are hiding behind.

Suppose, in advancing daycare, the government wants to encourage a higher participation in the workforce of mothers with small children, for the following reasons: Given stagnant real incomes and the high cost of living, they (politicians) are convinced that most young families basically need two incomes; they want woman to participate more equally in the political economy; they prefer that single mothers work rather than live on welfare, and they believe that jobs, and especially good careers are a force for woman's liberation; they may harbor a suspicion that on the whole children are better socialized, and less subject to abuse, in well-run daycare than in the average poor home; and finally, seeing the daily struggle many parents have over daycare problems, they feel the government should do something to help, and they believe direct involvement through operation of more government-sponsored daycare centers will not only provide more places, but will assure a good standard. And perhaps being socialists, they have a deep faith in government stewardship as less selfish than profit motives in the private sector.

Any "conservative" in the audience will have no problem raising a serious, "I see, but . . ." on every point. It will suffice to take just one to dramatize the clash of ideologies. The conservative replies: It is increased government causing high taxes and high deficits, which, making our economy less competitive, undermines growth in family incomes, causes unemployment and hence entices more mothers to work. Help mothers stay home by lowering family taxes, which is, if you will, a subsidy to the family home instead of encouraging the need for daycare.

Sensible people on both sides of such surface debates will concede, "It depends on cases, there are so many variables at play, ultimately

22. What is forgotten is that these special interests are often perfectly legitimate and desperately need to be heard in the hurly-burly of the HTX government arena.

one cannot simply generalize." But once an issue arises in the public consciousness—first crystallizing in the consciousness of the opinion makers, then through media distribution—it becomes an issue of public policy decision. And so it must and will be debated, if not down to fundamental principle, at least on a level several removes into abstraction from the concrete reality of *this* couple deciding what they will do this year with their little children.

The need to analyze one's way through such a maze of abstracted interacting processes and structures requires the managerial mind-sets of the HTX. Charlemagne, with the archives of the Holy Roman Empire on the back of a horse, did not think this way. I would propose, despite the difference in vagueness of goals, there is not much difference in the mentality underlying government "social engineering," on the one hand, and that of managers "reenginering" a company, on the other. Both seem to be molded by the same being, the illumination of the world in a way that has much to do with the fact that governments, like corporations, now have responsibility for such huge masses of people; that they receive avalanches of data, organized by typical concepts of contemporary social and political scientific ideologies; that they interpret these with a mentality of mass engineering, a mix of a kind of analysis and a will-to-mold on a grand scale, and the HTX provides the conceptual and institutional resources necessary: massive funds that can be borrowed, bureaucracies that can put in place and oversee multitudes of programs, huge universities that can train the necessary professional personnel; electronic communications and information processing, without which neither the data mass nor the colossal programs could easily be manipulated; mass media which can also manipulate—in this case "the public"—into an acceptance of such schemes, or into opposition to them. All these factors interact and reinforce one another, even as they reveal tensions within the system.

The resulting overly managed and yet still partially unmanaged behemoth leaves everyone feeling alienated: the client complains of a faceless, uncaring bureaucracy delivering too little too late; the social worker complains of underfunding, blunt funding cuts, insensitive higher levels in "the system" and blames "the public" for not wanting to be taxed to oblivion to provide more; the taxpayer finds the system wasteful (as do some of those within the system), with too much being creamed off by the bureaucrats and social workers, and he finds his taxes too high but feels powerless to change anything. As a physician friend says, as he contemplates the Ministry of Health trying to micromanage medicine, "We are caught in a whole network of tyrannies."

And yet unmanageable as it all seems, it would once again be "spoilt brattish" to fail to acknowledge the massive accomplishments of these clumsy systems in all the advanced HTX countries: universal free education, unprecedentedly good health care, universal pensions, considerable help to the poor. Much of this has proven humanizing, whatever the instrumentalization of the system. And, yes, "things do change" too, eventually, on the vast scale of interwoven bureaucracies. Recently in Canada we have seen a titanic struggle within the cabinet of a Liberal government, provoked by the warnings issued by the functionaries of international organizations—IMF, OECD, international banks—that the financial markets were going to punish Canada, if the governments at all levels did not address the unprecedented budget deficits more energetically.[23] The government senses that the electorate will tolerate no more significant tax increases, and yet tremendous resistance is raised by "interest groups" (and the "left wing" in cabinet, both aided and abetted by the media) to each and every proposed cut in programs. The government's strategy is to disguise real cuts behind rhetoric of "delivering better care through greater efficiency." At this writing, no proposal at all has been forthcoming regarding the two most expensive programs, universal health care and old age assistance.[24]

Cherchez le Soi Derrière L'Affaire

To be able to see these structural realities and processes of change and conflicting pressures within them, the analyst must have the self-knowledge and the emotional discipline to keep separate what he personally wants to see happen from what he observes to be actually going on.

Such a "scientific" stand is good HTX behavior. But its roots go deeper than HTX being, rejoining fundamental occidental being in its scientific roots chez the Greeks. In searching for the "Good" in any situation, one is really seeking being itself as *transcendentally* desirable, being (*esse*) as that toward which the will naturally moves. Aristotle al-

23. These have produced government debts now equal to 100 percent of GDP. Compare with Canada's almost all-absorbing trading partner, with an economy 10 times its own, the U.S.: government debts equal to 64 percent, in France they are 42 percent, in Belgium, 138 percent! Such comparisons are vital in the HTX for getting some sense of where one is on this vast scale.

24. More recently, there was a trial balloon suggesting pushing back retirement to 67, but only for those now under 50. This will have no immediate effect on solving the deficit crisis.

ready informed us, one cannot get beyond oneself to know the Other—the good that transcends the selfish ego—without a mature knowledge of, and hold on, oneself. The "selfish" person is not interested, really, in anything but himself, which, paradoxically leads to his lacking self-knowledge. He is not really interested indeed in his *authentic* self, but in "working out his feelings." One can come to know oneself only in situation with, and in contrast and comparison with, the Other, in whom one had then better be interested. In the selfish person, what he wants gets in the way of his judgments about what really is; he mixes them up.

Right there is the root of the temptation to stay within a manageable pseudoreality, a fantasy world of one's own, which, of course, is "a house of cards." When we see how difficult it is to cope with a reality that is so vast and hence is mediated to us through hierarchies of abstraction, and when we consider how much freedom we have in building conceptual models and in imitating the real through the construction of virtual reality experience machines, we can understand that selfishness installs at the base a temptation to forego the hard work of thinking required for any good hold on the world.

Seizing hold of the truth that human nature is at the base of the HTX as much as it is at the base of the being of any other epoch remains the key to learning to survive and even prosper in this "virtual reality" world of ours. So in the next and concluding chapter, I shall return to that base, what Nietzsche called *"die menschlich, alzu menschlich"*: "the human, all too human."

Chapter 6

Gigantomachia
The Clash of Anthropologies

HTX Builds on Human Nature

The nature-culture-civilization-HTX hierarchy is paralleled by a hierarchy of laws: custom (Hegel's *Sittlichkeit*), common law-statutory law, administrative regulation-Internet protocols.

Table 2

HTX	internet protocols
civilization	statutory law and administrative rules
culture	common law
human nature	custom (*Sittlichkeit*)
nature	structures of nature

I point this out by way of reinforcing the truth of the central role of symbol as expressive of, and participating in, the structures of society. The HTX reality, and our place in it, can only be understood in terms of man's convictions about himself. These are found in every kind of expression, from everyday banalities to the most exalted tomes of theological teaching. Daily fashion influences our lives, but so too do the theological dimensions of the various anthropologies driving the great traditions. Later in this chapter I shall examine the profound effect of this split in anthropological vision on the HTX and on what remains of the civilizations it is transforming.

In the hierarchy of social realities from nature to HTX, a dialectic back-and-forth of reciprocal influences is evident not only between adjacent levels (e.g., HTX with civilization), but also from top to bottom (HTX with nature). Cultures provide the "stuff" of civilizations, but reciprocally, the civilizational *Sein* partly molds the cultures. This being said, it is important to bear in mind that whatever appearance of freedom there is at the social, cultural, civilizational, and HTX levels in fact is entirely owed to the exercise of freedom by the individual human agents, including their response to both the natural realities of their bodies and to the "supernatural" call of the Transcendent. So, understanding the dialectic between human nature and every other level in the hierarchy is vital for grasping what is happening to the world. I invite you now to a moment of reflection on that human freedom which affects all—even nature itself, as man transforms it through his work, and I shall confront directly what I believe to be the most fundamental and significant clash of freedoms in the world today.

From top to bottom of these hierarchies the dialectics are producing strange interpenetrations. For example, HTX communications standardizes in many ways, but at the same time, by bringing us into greater intimacy with thousands of cultures, it uncovers a babel, close to nature (different languages, and different symbols). At the same time, paradoxically, English is spreading all over the globe as the *lingua anglia* of the HTX, and yet this does not guarantee that Iraqi technocrats speaking English in Baghdad understand a single thing about the traditions of English democracy.[1]

Moreover, in the face of the foreign onslaught, many groups are pulling their ethnicity around them like a cloak as protection against the winds of change. At the very time they are being drawn back to nature by renewed interest in their ethnicity, and being pulled up into the HTX web, many individuals and communities are deepening (inside their souls) their spiritual lives, seeking to be grounded in the Transcendent, allowing the Divine to reach down into them to show them "the Way." That this may lead to outbursts of Muslim fundamentalism, the schmarming of New Age neopagan cults, evangelical assaults on old religious cultures, and outbursts of private revelation (the Virgin Mary is showing up in the darndest places!) should not distract from appreciating the positive effects of the more responsible and profound spiritual awakenings.

The Three-Way Pull

Many denizens of the HTX are being pulled three ways simultaneously: down into the tribe, up into an HTX that then spreads them hor-

---

1. Think of all the dialects we can now discover being spoken somewhere in the world, even though many are being lost to more imperious penetration of the unifying national language (e.g., French or Hoch Deutsch). Still, *each person speaks his own language,* at least to some degree. I have to learn to pick up all the nuances of a friend or a competitor in negotiation.

izontally across a planetary world, and out into divine Transcendence. This last can be a very complex movement, leading at once back to God as Origin, forward to God as End and Fulfillment, up to Him as supreme freedom, and down into the depths of the soul where religious people believe He is abiding *interior intimo meo.*[2] As we shall see, this holds implications for the tensions established by all tribal, HTX, and religious movements stirring mankind, quite amazing if one mistakenly believes the HTX epoch is all efficiency and control.

A similar paradox in the media: at once a growing global concentration in the hands of a few giant media corporations at the same time as desktop publishing allows every little group a printed voice, and a TV diversification occurs: 500 channels are becoming available from "Death Star," and the Internet provides free access to more information and more entertainment, and permits wide expression of individual views, most, of course, somewhat drowned out in the babble, but also rapid formation of action groups. Chinese students in San Francisco and New York, armed with fax machines, made the mighty government of China tremble. Old industrial era means were used to crush the movement at Tiananmen Square. For the moment the tank was mightier than the fax. All of this is relevant to the proliferation of cults and "do it yourself" Christian free-lance churches, or as evangelicals see it, "Let the Holy Spirit do it" without most of the traditional structures Catholics believe were given by Christ. (The analogy between the erosion of "multinational corporations" by small, nimble competitors made possible by HTX communications and the inroads evangelicals are making on traditional mainline Protestant and Catholic populations is suggestive, but also treacherous. While you are thinking about it, remember that some of the international giant corporations are showing surprising vigor.)

The aggressive "tamper-with-everything" aspect of the HTX makes it difficult to take much for granted anymore, we are presented with so much choice, we can alter so much in nature now, our very status as human beings comes into question. Everything has to be argued for.

However permeated by HTX-formed desires, the agents who make the systems run are still subject at a deeper level to demands springing from their old civilization and cultures, and from primordial human desires from sources older than civilization, and some transcending all cultures and civilizations.

---

2. For a full elaboration of the interplay of these "ultimate structures," see *Being and Truth,* chap. 10.

Most of the demands that are made on us are not encoded in the genes, although our responses are of course in part genetically molded. Most of these demands are widely embedded in social structures in which we play roles into which we have been inculturated, interacting with other role players in a host of little daily dramas. The hard wired reality of the human players is of course foundational, but it gets molded by the social structures. Every *cosmion* into which we are pulled unfolds dynamically through a constant play of these little dramatic encounters between free individuals cast in various and usually typical roles.

Just how deep the more recent HTX social realities can penetrate the individual, transforming primitive layers, social and even genetic, we do not know. Many living in our HTX countries still go about their business in a largely industrial, and some in an industrial-agricultural mode of life, and many immigrant shopkeeper families in a preindustrial, even Confucian mode, despite the incursions into their thinking by the HTX symbol spinners. Just how far all enculturated social dimensions impinge on the concrete freedom of the individual remains a central issue in the social and psychological sciences. Whether, following prophets like Marvin Minsky, we shall want to change human nature at its very foundations through genetic manipulation, and how "designer genes" will alter human freedom are questions with which scientific-technological progress is already confronting us.[3] (Minsky entertains few doubts about our eventually being able to manipulate genes with almost no limits. He does admit, though, that this raises problems about who we [or, more accurately, who our "brain children"] will be, and about who we *ought* to want to become). Minsky is right:

> It is time that we start to think about our new emerging identities. We can begin to design systems based on inventive kinds of "unnatural selection" that can also advance plans and goals and can exploit the inheritance of acquired characteristics. It took a century for evolutionists to train themselves to avoid such ideas—biologists call them

3. "The more we learn about our brains, the more ways we will find to improve them. Each brain has hundreds of specialized regions. We know only a little about what each one does or how it does it, but as soon as we find out how any one part works, researchers will try to devise ways to extend that part's capacity . . . we will try to connect (these new inventions) to our brains, perhaps through millions of microscopic electrode inserted into the great nerve bundle called the corpus collusum, the largest database in the brain . . . In the end, we will find ways to replace every part of the body and brain and thus repair all the defects and injuries that make our lives so brief" (Marvin Minsky, "Will Robots Inherit the Earth?" in *Scientific American* [Oct. 1994], 111).

"teleological" and "Lamarckian"—but now we may have to change those rules.[4]

In the good ole HTX, we, not God, decide what *we want* living things to be. Still we might indulge in a little gratitude to "God" or "Nature" for having provided us with the building materials and the intelligence to do this. Minsky tantalizingly suggests that elements from "the ancient wisdom," which he admits retain much validity, may give us grounds not only for deciding what we want to be but even for hesitating about just how far to plunge into the Brave New World. (After all, man has possessed since Adam and Eve many capabilities to do things he has known he *ought* not do, such as cleverly deceive, manipulate, seduce, brutally kill a bothersome grandmother, covet his neighbor's wife—the list is long; now we can add, "cull out and destroy elements of the population we deem racially inferior, and destroy entire populations with biological weapons, even terminate all human life with a sufficient [and efficient—low cost per death] nuclear exchange.")

But have no doubt: the "managerial" mind-set activists will find it irresistible to avoid playing around with designer genes. How wonderful to engineer brainier children! (Of course styles will change, and one will be able to identify which *fashion* was prevalent when you were engineered.)

The *Economist* comments,

> What was once unique to genes is now in humanity's grip. That grip could soon have all the power that has at times been attributed to genes, and more. The same intelligence will be able to shape the gene and the environment, which between them make all organisms what they are. The control of biological information on that scale—of the raw data and the way that it is processed—means the control of biology, of life itself.
>
> The power of the gene has been revealed as partial and counteractable; the awesome biological power of humanity is still only embryonic. This interregnum is a good time to start making decisions about how best to effect the handover of power from nature to man. Those decisions will be difficult. But the general principle to help and guide them can be identified now: respect for autonomy, respect for variety, respect for equality.[5]

Upon reading this, my wife commented, "God help us! Do you see in everyday life now much respect for autonomy, variety, or equality?" I

4. Ibid.
5. *Economist,* February 23, 1995, survey section, p. 18.

would add that "the power of the gene" is being exaggerated here, if I am right in thinking that most of the molding of the character and personality takes place on the social, not the genetic level. (I shall try to justify anecdotally this faith statement affirming more nurture than nature later in this chapter.) Despite a million years of free activity in social interaction, the natural foundations in man remain almost entirely intact. Gene manipulation, however, could change that, perhaps by tampering with that in the biophysical makeup of man which provides the physical foundation for the spiritual act of freedom. Grave brain disease can shut down all power of free choice. So can deliberately designed quasi-human robotization to produce slave bodies.

"Natural Faith" and Worlds within Worlds and World Everlasting. Amen

We are encountering here, in this difference between the "if you can do it, do it" set and the "be careful, we are not God" folks, one of those basic facts about human beings, which even gene manipulation will not change, so long as man remains a cognitive being: The way a person has lived out and carries forward today's HTX demands is fundamentally affected by, and in turn affects, his "natural faith." Even the gene designer would proceed to design his children in the light of his own natural faith.

By "natural faith" I mean how a person thinks "it stands with being"—the basic beliefs that provide the big context of his worlds and govern his courses of action. One's natural faith never ceases developing, although basic planks in its platform may remain little altered from childhood to old age. People's differing natural faiths govern the big distinctions between orientations, at the base of the tensions in society, as well as the more regional contexts of the many *cosmiota* in which he plays roles—family, folk, religion, profession, clubs, and hobbies.[6]

The absolute pluralism of natural faiths founding individual worlds is, of course, only half the story. The other half, which, I believe, redeems the project of truth, is that despite differences in faiths, real persons, and things and real relations between them *presence* in the time-space we open to them, and challenge our assumptions, other-

6. See *Tradition and Authenticity,* 23–24, 33, 39, 132–33; and *Being and Truth,* 7–12. The natural faith of the activist-futuristic-scientific world of the Minskys has little in common with that of the Trappist monks in Genesee, New York. Members of both groups would agree that they are in the same country and the same century but in a very different kind of *present.*

wise no one would ever learn anything. Tensions within the individual, pulled by his basic human nature and the roles of the different worlds to which he belongs, the strains between individuals and between different groups, provide a competition of ideas and desires, affecting how our lives drive forward. But they are also obviously the source of confusions, even leading to psychic disturbance and to war. On the other hand, the broad "common sense" horizons one shares with all other human beings, and the narrower ones shared with all within one's civilization (and now within the even larger HTX), tend to be "taken for granted." They are a source of stability, forming the core of the everyday world within which all the particular little worlds meet. There is a downside to that stability: it can be, within a culture or civilization, a millennia perdurance in error—"Come on, Oddball, everybody knows the world is flat." A less ambiguous stability is provided, on the other hand, by the real endurance of persons, cultural objects, the things of nature, and many of the ongoing real relations between them.

These broad common assumptions and the widely used symbols in which they get expressed partially orchestrate vast shared mind-sets, and the cultural objects produced as a result of such ideas. Because these worlds of notions are so all-permeating and so hard even to think about explicitly, they are difficult to change deliberately, even if some begin to worry about aspects of them leading society in a dangerous direction, such as an ugly racism, or excesses of workaholism, or extreme technological manipulation of life. But some fairly basic principles will shift over time; "everybody" in Western society now knows and believes that the brook and the woods are not inhabited with gods and leprechauns ("Wiccas," I assume, are still beyond the pale of common sense), that the sun is not the center of the universe, and that slavery is immoral. On the other hand, not everyone yet believes that profit is vital. But where the agreements of "common sense" end, serious disagreements start between inhabitants of different little worlds.

HTX thinking has penetrated the natural faith of everyone in developed countries to some significant degree, becoming part of unquestioned "common sense." Each person has his own style of living in it; and in different social ways—the distinctive faiths of particular groups and communities form worlds within the common human and the vast HTX world that are as different as Muslim ways of living out the HTX are distinct from atheist ways. There are higher management ways of living the HTX, different from organized labor styles.

The hierarchy of worlds within worlds is clear: At the base is
—the little world of the individual, in whom all worlds meet; next,
—*cosmiota* are made up of shared personal-individual worlds;
   these include the many different cultures;
—larger *cosmiota* embrace smaller ones; these include the horizons
   of the distinctive civilizations
—the epochal HTX world encloses most of these smaller worlds,[7]
and
—the largest world of all, "THE world, the horizons of interpretation based in fundamental human nature, embraces even the
   HTX, and goes beyond the accepted wisdom—the "common
   sense"—of whole civilizations to be found contributing of the
   sense of every human project.
Five kinds of "enfolding" of worlds by worlds can be distinguished:

1. the personal world projected by the individual "center of awareness and initiative" enfolds all other possible worlds.[8]

2. The intersubjective planetary epochal world ("the HTX," or THE
world) enfolds all other worlds in its conceptual reach and vast inter-
subjectivity.[9]

3. Very similar, but on a lesser scale than the epochal world, are all
the *cosmiota,* the little worlds, all of which involve fewer phenomena
and kinds of phenomena than the whole HTX, and fewer subjects ac-
cessing any one of them from their individual perspectives.[10]

4. The cosmos, the real cosmos, not anyone's notions about it, is not
a world in the phenomenological sense but a dynamic whole in which
are to be found interrelated all persons, all things, indeed all relations,
even the "finite pole" of relations with God, who, if it is true what Jews,
Muslims and Christians believe, transcends the cosmos He created. So
in a nonnotional real sense the cosmos enfolds all, so to speak, physi-

---

7. We can live intensely a large part of our lives in one or another of these little
worlds—the all-absorbing world of a Trappist monk, or that of the CEO who eats,
drinks, and breathes Motorola. Well, not quite *all*-absorbing, for even the monk
will tell you how "plugged in to the outside world" he is, and even the CEO shares
in the "common sense worlds" of his class and civilization. He may even have a
little space for his family. The pull on the individual of these different worlds he in-
habits, with differing intensities, is one of the most basic of the many causes of
stress in HTX life.

8. It enfolds subjectively but also ontologically, in the sense of *Sein* (the illumination of all that is through the human horizon-opening interpreting existent, the
Da-sein)

9. Here it is intersubjectively and again ontologically—in the sense of *Sein.*

10. Each such little intersubjective world has its own *Sein,* e.g., the being of the
world of commercial sports.

cally, including humans' participation in, knowledge of, and perhaps a love relationship with, the transcendent Source.[11]

5. We need a symbol to stand for the most global of englobings, the belonging together in the unity of being of everything "objective," whether known by humans or not, and everything "subjective," whether referential in a true way to what transcends consciousness or sickest of fantasies, and all relations between them, including the possibility of a relationship with the transcendent Source. (He would obviously be a being quite unlike all the beings revealed in our direct experience.)

There is no need to create a barbaric neologism. The term has existed since the beginning of philosophy, those first efforts by the Greeks to think the sense of all reality: *being,* understood as the analogical notion embracing everything and all relations between things. (See the addendum at the end of this chapter for some elaboration of the notion of "the analogy of being.")[12]

The Basic Conflict in the HTX's Homeland, Western Civilization

Whatever the heights to which philosophical speculation and recovery of the great tradition of "the analogy of being" might take us, we remain solidly implanted in the present HTX world. To deal seriously with the resulting huge HTX reality will require a series of sectorial analyses, focusing some on different parts of the world (e.g., the Japanese form of HTX society and its impact on the world, Russia's, North America's, Western Europe's); other monographs could study each of the major traditions—Islam and the HTX, Christianity and the HTX, yet others on various segments of life, such as HTX medicine, HTX entertainment, and HTX education. This would have to be the work of specialists, but who share something of the vision of the being of the HTX such as we are trying to evoke it here.

But I do not want to conclude the present modest introduction with-

---

11. My knowledge of all this cosmic reality is, of course notional, it projects *Sein,* as does the intersubjective community, of, say, cosmologists, and as notions themselves enjoy only *inesse.* But what I and the astrophysicists express through these notions is true if our imaginations rightly represent the state of affairs that transcends my ability to know but a little bit of it—enough to know there is much more, and to be cautious about what I affirm of it, even though I realize it may well be true as far as it goes. No one understands gravity very well, but it is a universal phenomenon affecting at least gross objects, that much we do know.

12. *Being and Truth,* 353, 356, elaborates this very special notion.

out launching a first reflection on what I believe to be *the* fundamental *ontological* conflict within that part of the HTX world that may still justifiably be called "Western civilization" (or what is left of it). I intend to show how that ontological split, buried deeply within clashing "natural faiths," affects each of us personally. My own natural faith veritably shouts "Facing up to this is vital for survival!"

I gave a big hint what it may be when we were discussing the genetic redesign of man: It is the competition between an interpretation of the human condition *respectful of the mystery* surrounding man's coming to be and hence careful about such projects and *a materialist natural faith,* in which it has been decided man is free to make himself into whatever he can imagine—"the receptive" versus "the activist."

A techie friend who is running a management seminar for the CIOs of four of Holland's largest firms tells me he convinced them, "You can no longer look to the past for guidance. It has become irrelevant. We are entirely about inventing a new future." Well, we shall see about that. Their firms, for instance, will continue to be staffed by the same old human beings who ran *Scrooge and Marley,* people who don't like to be "downsized," and some who do not relish chopping off loyal employees. (The morning Chase Manhattan Bank opened its startling new branch at Fifth Avenue and Forty-third Street—the first Skidmore, Owens, and Merrill glass cube—Gordon Bunshaft, the principal architect, stood with the chairman of the bank watching the employees stream in with their lunch sacks from Queens, Brooklyn, and the Bronx. Once they were settled in and the doors were ready to be opened, the architect asked the chairman, "How do you like the building now?" "It is beautiful, wonderful . . . except for the one important thing you forgot." "What is that?" asked a worried Bunshaft. "You forgot to redesign the employees.")

To be sure, any hope of finding one's way through the human confusion of the HTX requires more than a retooling of "the ancient wisdom," rather more something like a rebuilding of wisdom from the ground up but with much still drawn from millennial experience—even Minsky sees that.

Who "we" are in all this is arguable: "we" are, for sure, a species, a fact that can breed a vague "humanism"; *continuity* can be found in the family (Confucianism), *togetherness* in the race (tribalism, including grotesque modern forms like Nazism); we certainly are bound to a nation (this can breed forms of nationalism), and all—humanity, family, race, nation—can be turned into idols, and used to deaden the pain of individual death, and (maybe) to stave off total meaninglessness.

But there is another certainty about "us"—as far as life in the cosmos is concerned, mankind will certainly come to an end. Barring aliens emitted from boob tubes, there are only two possible ways for this to happen: either we eradicate ourselves by means of our own invention (that new possibility is an unwelcome very recent fruit of HTX inventiveness), or the fragile conditions that permit *Gaia* to sustain us change too much to support such a complex life form. Further down the line comes a sure end, not just to man but to everything, even if we flee the death of this planet to a nearby favorable planet some light-years away (I heard a NASA scientists seriously propose this in a public forum.) The irreversible enemy: *entropy,* with no "information" remaining in the ruins of a dead and motionless universe. The "immortality" of family, tribe or nation is temporary. *Nothingness* awaits just over the horizon, and it can seep into present consciousness as the deadly gas of Meaninglessness.

For many "humanists" it seems sufficient for humans, who are by nature social, to work at maximizing opportunity for all. (HTX form: "good management in the mode of enlightened self-interest.") A certain noble altruism pays, they believe, and many devote themselves to making possible a better world for all. Marxists even take a long view and are prepared to sacrifice much in the present generation to make life better for future generations.

I suppose clear-sighted explicit cynics are relatively rare, those who pooh-pooh altruism.[13] Selfishness, on the other hand, is obviously not rare, among theists and atheists alike. This leads to behavior, regardless of what the person may loudly profess, that we can justifiably call "practical atheism."[14]

In chapter 4 I raised the question how the different degrees of corruption and criminality in various countries and segments of HTX society affect the system as a whole. We took note of the contention of Davidow and Malone that, as manufacturers and service providers work with customers and suppliers to reduce delivery times and increase flexibility, more *trust* on the part of all, especially the sharing of

13. The Frank Lloyd Wrights and Ayn Rands are exceptional and merit consideration on their own.

14. This is because every selfish act implies there is no purpose higher than one's own gratification, often indeed one's *immediate* gratification. Hence not only do the demands of God take no real place, even pressing communal demands are to be massaged in one's own selfish interest.

proprietary information and confidence in passing some decisions to the other levels becomes "the defining feature of the virtual corporation."[15] *But if there is danger of the HTX being compromised by moral rotting out of the base in certain key societies, the question then ought to be whether the future of the whole HTX interweaving of systems might not be in danger,* with massive "dislocations" coming upon us rapidly.

Recently a lawyer friend told me that his clients seeking to do business in Mexico were appalled by the Mexicans' alleged lack of respect for contracts, and the "total corruption of the judicial system" to the point that they were giving up on any effort to do business there. He also recounted how *Playboy* abandoned the Russian scene after the company received a box containing the head of its Russian partner, because HQ in America refused to pay the Mafia protection money. (It has since returned!) Another friend with large manufacturing operations says he has no trouble with the Mafia, because he hired the top security service—"A-Alpha," made up of ex-KGB agents!

Now to these anecdotal indications come the appalling revelations of fundamental corruption in the Orient brought to the glare of publicity by the financial collapse termed "the Asian flu." I wonder to what extent the countries with old traditions of rule of law and considerable segments of the leadership still influenced by a civic sense are able to resist contamination by countries with whom they are almost compelled to do business (e.g., China and Indonesia) where the rule of law is feeble and corruption well installed.

The *Economist* estimates at $200 billion the money available to criminal elements in the USA from the drug trade alone, a sum exceeding the surplus thrown off by the entire Japanese economy. How many people do you know who could resist a million-dollar bribe?

So here is perhaps the supreme HTX paradox to be faced: how long can we go in opposite directions, increasing need for confidence and trust in the "virtual corporation," while, at the same time, the Mafia mentality is spreading from China's Communist party and Mexico's PRI to the money-laundering operations of Little Rock?

The Utilitarian Nature of Many HTX Relationships

I can hear the reader sigh: So what else is new? We acknowledge this evidence: "'Users' you will have with you always."

But there is a development in the accelerating HTX that poses a real

---

15. Davidow and Malone, *Virtual Corporation,* 9.

and constant danger of a further "dehumanization" that turns the "user" phenomenon, present since Cain slew Abel, into a potential terminal threat to man: The *essentially* utilitarian nature of so many relationships within the HTX.

This has crept up on us so steadily for so long, we start to think that mankind has always related that way. Not so. This utilitarian way of relating arises sometimes because of the remoteness of the relationships, and the fact that we think about them through hierarchies of abstractions. A manager recently promoted to manufacturing general manager told me that when she became "a manager of managers," she thought she should change from the much more personal relationship to subordinates she had before to a more impersonal style, for fear that managers, "who are always trying to psych out what you want as you pass down some rather vague directives from on high," might take advantage of friendship. Other times it is the narrowness of the relationship—I just want a transfer from the bus driver, which could just as well come from a machine. (In the village, you don't even pick up firewood without chatting up a storm.) The condensation of time also contributes, for interacting "humanly" (which means considerately, patiently, quietly) with another requires time, and time is what we ain't got! This seems in flagrant contradiction to what we are being told about the "virtual corporation" requiring workers to take more responsibility and initiative, and about the need for trust, and for loyalty to subordinates.

Somehow this all recalls a rather famous comment on genuine human relations from St. Paul's first letter to the Corinthians (13:4-7):

> Love is patient and kind. Love envies no one, is never boastful, never conceited, never rude; love is never selfish, never quick to take offense. Love keeps no score of wrongs, takes no pleasure in the sins of others, but delights in the truth. There is nothing love cannot face; there is no limit to its faith, its hope, its endurance.

Some Dilemmas from the Real World

It will prove helpful, once again, to descend into the hurly-burly of everyday reality to reflect on some anecdotal examples. A few anecdotes do not substitute for social science. But at least by way of them I may contribute to the raising of some of the right questions.

I have just had discussions about the human implications of some of these restraining realities as they affect the workers of my small company which manufactures wire assemblies. The foreman was concerned, not so much about the low starting pay ($7.50 Canadian), but with the final ceiling ($9.20 Canadian). Most of our original workers have remained devoted and productive. He had a second concern: the monotony of the work. His assistant, under the pressure of keeping the work flowing, finds it difficult to think to switch people around several times during the day so that they do not get too bored, and he doubts that production would be much improved if he did. It was obvious that the foreman thinks of these thirty people as human beings, he cares whether they are happy or not, beyond the effects of their moods on production. But he really does not have much time to dwell on anything not essential to production.

It is amazing with how many details they have to deal: our main customers order in smallish lots some 130 different models of wire assemblies, some of which are quite complicated, and there are some 30 customers. The manager talked with more enthusiasm about the Vietnamese workers' superiority, not only in dexterity, but in paying attention and so catching mistakes made in earlier steps of the production, than he did about the possibility of keeping everyone from getting bored to tears. I asked if the workers often complain, and it seems that they do not, probably because they understand the pressures and appreciate that the foremen do care and are doing the best they can, so they try not to be disagreeable.

The president, worried because the company had accumulated too much debt over the last recession, and concerned by just how thin are our margins of competitiveness, was asked about giving the older workers a better deal, and he responded that we simply cannot and remain viable. When asked whether, once we pay down the bank debt and begin giving the shareholders a return on the loans they have made to the company, we might start a profit-sharing scheme, he replied favorably. But I could see that was sufficiently remote as of now as to strike him as merely theoretical. But we do offer good benefits and always give Christmas bonuses.

All this led to a discussion of the insistence in the Catholic Church's social teaching that managers pay "a living wage," and of the difficulties of determining concretely what that might be. In the present local labor market, our wage is obviously desirable—we have no difficulty keeping excellent people—if not "fair" (in the sense that raising a family on what we pay is next to impossible). The Church's social teaching displays little mercy for owners, but for the record, after five years, we have yet to see any return. I added that the real buying power of our underpaid workers places them in the top 5 percent in the world, although life for them here would be judged hard by Canadian standards.

Here is another example, to emphasize that the towering HTX structures reside on a fragile "human, all too human" base:

Is It a Cop-out to Refuse to Go up the Ladder?

A successful manager in a corporation with 22,000 employees told me that his boss was puzzled that this superintendent was working his engineers so hard but still had been rated by all with the highest marks. The manager explained that he had set for each engineer a clear area of responsibility with precise goals and then given him control of that part of the project, including budget and administration of assistants. Each then derived satisfaction from the accomplishment of his little section. At the same time, each understood the essential contribution his project was making to the mandate of the division. This mature relationship was working.

But it worked because it was *immediate,* the responsible authority could work out directly with the engineer the challenge and the responsibilities. Also it helps that they are doing technical tasks that demand accomplishments you can see which cannot be disguised by hype. The particular company belongs more in the industrial than the HTX phase two era.

The young superintendent, whom some had told "You ought to be the president of this company," mused about his fears in moving up. The next level of management he sees as largely superfluous—the general managers neither do anything nor do they make the big strategic decisions, which lie with top management. He believes that layer will be eliminated as the hierarchy is "flattened." "The general managers sit in meetings all day and never get anywhere near the technical things this company is all about and which I, as an engineer, enjoy, it keeps my mind alive to help solve tough technical problems, to get away for awhile from politics." But a further matter adds to his worry about moving up. "At the higher level, it is game playing, much hangs on negotiations. In negotiations you can never tell the whole truth. You might not out and out lie, but you exaggerate. For instance, you make a bid, saying 'Oh sure, we can handle that within the time frame with the manpower we already have,' while thinking to yourself, 'How am I ever going to do that?' and just deciding, 'Oh well, that's for tomorrow to find some solution; for today, I have got to win the business.'" He considers himself perfectly able to play that kind of game, but doing so offends his moral sense. "I have my doubts that I want power at that price"—the sacrifice of self-esteem, and remoteness from technical issues that increase efficiency and quality, "rather than just winning political games." (He is however aware of how the outcome of those political games can "empower" the technicians to do something worthwhile, or can gut years of valid effort). He added that in a recent shakeup of his company, he took note of how many "successful" higher managers suddenly found themselves out on the street. Finally, he added that he puts a high priority on being a serious husband and father of three, and that in his present position "I don't have to shortchange my family." Wisely he is struggling to balance two worlds.

Since first penning this example, the superintendent yielded to pressure and became a general manager, with orders to prepare and manage a budget of $500 million over five years. Most recently he was asked to let his name stand for a vice presidency. He flatly refused. His predictions have come true: he is caught in endless negotiations, he is remote from the action, and he is getting home at 8 P.M.

It is interesting to hear such reflections from a super-skilled HTX man, someone who reflects both as a human being and like a manager about what he is doing in life and what he wants to give his family. I have known a number of successful young managers who have deliberately avoided moving up higher in the organization when they esteemed that it was better *humanly* to stay where they were. This of course leaves the way clear for those with a different vision. Again: Poor strategic decisions at the top can vitiate solid accomplishment farther down. Is that not a reason why the conscientious, talented family man owes it to society to move up? But does not the individual have the responsibility to himself and to his family to stay where he will be most fulfilled as a complete human being? He can always hope that there will be other good people who will allow themselves to rise to the top, who may either have no family to worry about or perhaps have found some magic solution to the absence-of-parent-as-model dilemma. So optimistically it can be imagined that not everyone "negotiating" in the stratosphere far from where the company produces will be making poor or mean-spirited judgment calls, nor sacrificing the children's well-being to that of the society.

I showed what I wrote above to the young manager in question. This is what he penciled in the margins of the manuscript:

> Actually, the irony is that the only people who can alter the reality of the company are the people who *do* the work, who create the prod-

ucts. They are the real power. "Power" is illusory. I have seen CEO after CEO fail to alter the company significantly, after being criticized by the vast majority of the staff. In the end you can't force people for very long to do things they do not agree with.

Postscript: In the interim, the inner disarray of this large company, once a leader in its industry, has become public knowledge. "Downsizing" games, badly played, have so gutted the morale of a once model company, its ability to produce a quality product has been compromised. How HTX this all is. These are new games, quite different from those played at lower levels or in stodgy old bureaucracies.

Playing the New HTX Game

What we have been encountering is these two anecdotes is a basic, you might even say banal human dilemma in the exercise of power. In every society since civilization began each person within the commanding classes has had to struggle within himself about where and how to position himself. This interior struggle is fed by the tension between what is demanded by "the game" and what the individual believes is expected of him by virtue of who he is, both as a human being (with the religious question of our destiny), and as *this person,* with this station in life, these talents, particular intimate relations with certain other persons, and hence these particular desires and possibilities. (I would rather not try to imagine the concern for the children shown by the power elites of, say, imperial Rome, feudal medieval Europe, or those at the court of Louis XIV.)

In the tribe, one's position is decided for him, and the children are raised by the whole village. Only since the industrial era have many of these issues become relevant even for the lowest classes. They have become more of an issue as possibilities multiply with the wealth of the society. Each person enjoys, and different kinds of society provide distinctive kinds and even a different degree of possibility in this mobility: much greater with the elite than with a slave or an unskilled worker, for whom the room to maneuver can approach zero, much higher in the HTX than even in the higher echelons of the pharaoh's administration, or anything Pascal would have known.

I see two differences distinguishing in this regard the advanced HTX from all other societies:

First, an unprecedentedly high degree of self-awareness is possible, in part because there is so much mobility, and, of course, so much formal education, causing many questions to be raised more clearly than in traditional societies, where much more is taken for granted. That relates to the second consideration:

The openness of the society, its "multiculturalism," and its dynamism produce an exceptional array of possibilities for those favored with the cultural background and the technical education to play with a good hand. Those in the so favored segment of society can position themselves with less risk than in preindustrial societies, where a misstep could cost you your life. Consequently, more ordinary people are forced into some degree of strategic thinking for themselves than ever before.

Given so much to think about and readily accessible resources for helping to do so, you would think a new seriousness would be evident. Instead we see the *trivialization of life* and the pervasive popular forms of *mindlessness* rampant in "consumer society," producing massive irresponsibility. To be sure, many are taking the situation seriously, reflecting critically on their position in the game, and are concerned with the common good of the society. But they appear overshadowed by a massive social effort at escape. What might still be done to redress a headlong flight into meaninglessness that can, unarrested, only lead to disaster?

The Problem of Imagining the Situation

What Can You See? What Do You Want to See?

Those with highly developed technical, professional, and managerial skills may enjoy a gift of *discernment,* the ability to cut through hype and disinformation and vague abstractions to form some reliable picture of what is actually going on in singularly complicated situations. With all that, many technically well trained game players are finding themselves stranded, for instance downsized at an age when they are not rehireable, or with young families they do not think it wise to uproot.

Here is a juncture where one's "natural faith" obviously becomes critical, how one has chosen to situate himself within the spectrum of "great choices" will determine what in the picture he wants to seek out and then seriously consider. Depending on his fundamental, often not carefully examined decisions about *what he believes important in life,* he will include or exclude entire realms of possible evidence about what the real world is, how it has become this way, what is its destiny, and how his life fits into that. Without being clear about his fundamental choice or aware of many of the implications of the path he has chosen, he will nevertheless long ago have joined a camp in the *Gigantomachia,* the clash of visions within the HTX that some are calling a great *Kulturkampf.*

Degrees of mindlessness exist in denizens of all the classic tendencies, in all visions and in every epoch. But in the HTX world, so much pours over us, so many little decisions have to be made (even for a sophisticated courtier of Louis XIV much was fixed by tradition, and he was "plugged into" a much narrower world than the humblest Net Surfer!), who has time—or the education, including the self-discipline—for long, reflective considerations?

But certain basic choices make mindlessness more inevitable. For instance, if what one believes important is just playing the day-to-day game, then he will let the agenda be written entirely by the segment of the HTX society where he happens to find himself, and he will have given himself little incentive to question its assumptions. For the rest, distraction and entertainment will keep him from much serious examination of the ultimate meaning of the life he is building overall. "Running life's errands" (Kierkegaard) is the formula for a shallow practical nihilism that leads only to the grave—probably of the society as well as of the individual.[16]

At the core of all this—the very soul of the individual's natural faith—are the attitudes (*Verhaltungen*) and character the individual brings to living out his life in each of the little worlds, the good habits he has built up that allow him to stick to a course he has set for himself, or the weak character that leaves him to waver in the face of every new challenge.[17]

The World as Imaginative Construction

Remember, whatever picture we build up of the world is *an imaginative construction,* narrow or vast, filled with symbols rich in content or poor. What we are *willing* to think about plays a big role in determining what we *can* imagine, which is further conditioned by what we have experienced, read about, or seen in art, photos, or television. There is a

---

16. Read Pope John Paul's descriptions of "the culture of death" in his *The Gospel of Life* and other writings.

17. The term *attitude* loses some of the sense of the German word, *Verhaltung,* used by Heidegger, which means literally "a way of holding (*halten*) oneself." The English word *comportment* is closer because it emphasizes action. "Attitude" is weighted on the cognitive side, but if we do not forget it means what finally determines what one does, we can use that term. The consistency of our most basic ongoing attitudes is rooted in the habits we have built up, which constitute our character.

circle here: what we are willing to think about is also conditioned by what we are *able* to imagine. In reading books, our imaginations are fed; sometimes vast world schemes are painted for us in high-flying conceptual language, perhaps underscored with rich images. But again, that to which we have been exposed affects what basically interests us, but that is also always a question of choice.

There is always the danger—especially real for artists and academics—that one's imaginative construction of "the world" will become an end in itself, detached from concern for objective reality—building it up will become the more absorbing game, substituting for the real game of life, out in a cosmic and intersubjective world where nature makes its demands and freedoms interact.

I can easily begin to take my imagined world for the place I want to inhabit, pushing out intrusions from the real world more and more, isolating myself in a dream world I try to control perfectly. If I stay as much as possible in a world of the imagination, I can introduce more controllable friendships of a nonintrusive kind, such as the "virtual friendships" one might strike up on the Internet, attaching whatever face to the incoming text I might care to, and disengaging myself from any more intercourse at the first moment the "Other" displeases me. (The spirit of this was caught by Roseanne offering advice on the *Late Show* with David Letterman, "Forget the institution of marriage; if you like a guy, live with him until he annoys you, then throw him out!" [Wild screams of delight from the audience.]) With superficial friends, in pubs so noisy all conversation is excluded, nothing very serious is likely to happen. If, on the other hand, one is committed to family, just as when one is active in a parish, he will be stuck with real people whom he would not always choose as friends, and that can be quite revealing.

Surviving versus Appreciating

A thoroughly instrumentalized humanity, while reverberating to political correctness and endless talk of rights, tends in fact to become less respectful of human life concretely, especially the "useless"—unwanted "fetuses" and expensive old people, not to speak of the severely handicapped who will be "disposable," as Hitler and Ceausescu thought they ought to be. Equally worrisome is the reticence in an increasingly nihilistic society to bring children into the world.

In this context I have puzzled over the question of why the kind of attitudes I have just been describing lead to a cult of sinister ugliness. Yesterday my wife showed me a page from a magazine that was advertising the latest young people's fashions. There was not one model who had not been made to be as troubling and off-putting as humanly (diabolically?) possible. Is it perhaps because beauty is a radiance of being that leads to the Source of the very possibility of all such radiance? Whatever radiantly *is* attracts us to come out of ourselves to contemplate it, which challenges the whole project of our absolute autonomy, hence the selfish person's (and crowd's) desire to destroy what will not let us (either as individual or as faceless crowd) alone to be our little narrow controlling selves.

Paradoxically, the mindless, while considering themselves absolutely free, at the same time want to be taken over (temporarily) by forces that control them. Why this passion to be swept up in the crowd, swaying and shouting to the jungle beats of a rock savagery performed on blasting HTX instruments by beings as ugly as the big moneymaking producers can fabricate?

There remain vital elements in our civilization that work against such tendencies. They are found within religious communities but also outside.

The Brutality of Change in the HTX and Its Inevitable Relativism Alter the Basic Human Situation

An inhuman brutality of change characterizes many segments of the HTX. One can scarcely imagine what it must be like, for instance, to try to keep up professionally with the progress in almost any area of "Information Technology." Images dance in our heads of modern monks in loud sportshirts working nineteen-hour days seven days a week in Silicon Valley (one wonders when they find time to tool around on their Harleys), with the refrain drumming in the background: "Your new product has eighteen months to recover R&D costs!" Companies that try to amortize their desktop computers by keeping them three years run a serious risk of becoming fatally uncompetitive, and that is not legend.

On the macroeconomic scale, we witness daily the effects of "volatility." The much commented on (and somewhat exaggerated) claim of a shift of power from sovereign national governments to financial markets, coupled with the explosion in world trade, recognized in, and furthered by international trade agreements, has in fact led to events like the 50 percent loss of value in three days of the Mexican peso in January 1995, followed by the dictating of Mexican economic policy by the U.S. administration and the IMF.[18] The Asian flu was first detected by fever in the money markets, but what was then exposed were the poor and corrupt policies of many of the affected governments. Strong governments, with sound economies, retain great influence over markets. (Careful, Mr. Greenspan, do not stress that last word too much, thirty-year bonds will plummet.)

The high unemployment one finds in Europe and Canada, especially among the young who are usually counted on to provide optimism, is troubling. For the moment only the United States has fought its way out of that slump, even Japan has succumbed. Falling education standards and special problems of employment among black youths in America (now improving) and Turks and North Africans in Germany and France add to the disquiet. I have witnessed the discouragement of ageing parents faced with the prospect of having no grandchildren.[19] And is this sense of discouragement not fed to an extent (little mentioned in the media until the American president, showing dynamic leadership, forced the issue onto the front pages of American and Canadian newspapers; the Europeans are too jaded to care anymore) by a general disgust at unmistakable lowering of moral standards, not just in sexual conduct and in marriage generally, but in petty thieving in companies, shoplifting, drug dealing, and the vulgarization—especially the sick violence—of the popular entertainment media? Increases in serious acts of gratuitous "terrorist" violence add to the disquiet.

Add to all this the increasingly obvious fact that big government welfare states cannot solve all problems. A large portion of the population of any highly developed (and aging) HTX society has come to depend on the state, and many of these people are faced, as governments

18. Imprudent economic policies, refusal to clean up politics, and failure to make industry more modern and competitive are more quickly and visibly punished. Abusive governments may be obliged to act more in the general interest.
There is the perception that this volatility makes the average family less secure. But statistics do not justify widespread anxiety. Real incomes have largely stagnated, but they are enormously high by historical standards. Job-hopping has been popular for some decades and, statistics suggest, has not recently worsened, although it may have become less voluntary. While a reduced demographic base may be unable to support the pension regimes and socialized medical care for the elderly, the old in these highly developed countries are mostly well off.
19. We know five couples all over fifty-five with a total of nineteen children who have among them a grand total of *three* grandchildren. I can assure you, they are depressed about this. It forecasts the end of a race!

"downsize," with the need to fend for themselves, and they are ill-equipped and ill-situated to do so: the elderly, the poorly educated, the mentally disturbed, single-parent families—this adds up to a mass. It is hard to convince the "liberals" that the state, already taxing the producers close to the limit, cannot simply borrow more money to support huge numbers of nonproducers, to police the society (incarcerating—at up to $60,000 a year per prisoner—an increasing portion of it), and to develop and maintain an exceedingly complex HTX infrastructure.

The internationalization of financial markets has placed the most visible and brutal limit on the ability of states to extend themselves and continue wasteful practices. These limits are in some ways more definitive and all-penetrating than the traditional limit placed on states by their neighboring enemies. Some gurus, like James Dale Davidson, go so far as to declare the social welfare democratic governments bankrupt and the national state dead! Liberals find both claims exaggerated, indeed they see the whole thing as a "conspiracy of the banks and the greedy corporations."

Greater citizen participation in the realities of politics may well be part of the answer. But the time dilemma this poses brings us right back to the question of the "HTX Multiple Personality Disorder" (HTXMPD) we have been raising throughout this reflection.

As the pluralism in the HTX, bringing an unprecedented competition of ideas, leads to debate about everything, undisciplined chattering about ill-thought-through ideas simply produces more relativism rather than commitment to the careful, thoughtful probing of data needed to establish the truth. For this reason, it becomes ever more urgent, first, to avoid the temptation to run away and wall up in a ghetto, and, second, to recall that this same wealth offers opportunity. *It is up to the mature individual to decide to preserve time-space for quiet reflection and for varied and profound friendships, where it is not just information that is shared, but heart—where the whole person is present.* Easier said than done.

Nurturing the Human Base

Trust demands loyalty. They both have something to do with human goodness, which is not basically driven by "efficiency." True efficiency, not a simplification achieved by dropping standards, is good when solving problems of "productivity." But human relations are productive of love on a different basis. One could put the point cynically by acknowledging that the HTX manager who is *really* efficient in the long run, and not just obsessed with the immediate bottom line, will integrate human nature into his equations, he will seek not just narrowly competent people but good people with whom to work, building relationships of trust and confidence, which cannot happen in the absence of all loyalty. A good manager must demand performance for the good of the whole organization, removing—as gently as possible—those who prove incapable or unwilling to perform. Goodness is basic, *sympathie* is grand, but we cannot avoid sacrificing also to the great god Efficiency, and His only begotten son, Profitability, for they alone allow survival in the HTX competitive world. That is a basic reality of the HTX being, as the Soviet Union found out the hard way.

That does not mean that dehumanization must inevitably triumph.

One's hopes in this regard are grounded in one's faith—how one believes it really stands with Being. These hopes, by guiding man's aspirations and actions, feed the cultures and clashes between cultures within the HTX. Yes, cultures, for, square within the Western HTX city, you will find, not just as remnants, vibrant cultures, little societies of people gathered in intact families, churches, synagogues, mosques, and religiously and humanistically inspired communities, societies devoted to caring for the disabled, the sick, the homeless. The cultures kept alive by these societies, and the traditions animating such communities, are, in the process, nurturing the human nature of the participants—givers and receivers alike—keeping alive roots in the soil of our common natural humanity, and in rich old traditions.

These cultures often have little to do with many HTX cultures, although participants quite inconsistently embrace elements of both nurturing cultures and destructive anticultures, the dehumanizing and the uglifying. Good people are often desensitized, so they do not see the extent to which the more nihilistic anticultures, those that in no way root people but rather destroy healthy kinds of belonging, are driving the dehumanization. *That struggle to keep alive the source of true human life in the midst of every effort of HTX distraction to obscure it and the nihilists to destroy it, is the dramatic core of the epoch dominated by HTX phase two.*

Our HTX advanced societies are risking fiscal-political-social meltdown. Consider the example of the largest and most dynamic HTX society, the U.S. Craig Karpel, in *The Retirement Myth,* warns that 65 percent of Americans have no savings for retirement. Average financial assets per household of those who have saved is less than $15,000, and the average pension will replace less than 20 percent of preretirement income. Peter Drucker recently predicted for the new generation a retirement age of seventy-nine! Personal bankruptcies are making new highs every year, right in the midst of the longest sustained economic upturn in recent times.[20] Savings rates have turned negative, as people count on their inflated stocks. Meanwhile, younger workers are enjoying far smaller wages than other postwar young workers. Raised to be materialists, they are not in a happy mood about seeing prospects for the good life sinking before their eyes. In the midst of all this, a recent poll showed only 4 percent of Americans favor cutting Medicare if that means old people would have to pay more, and Medicare costs are soaring out of control: Luddites ahead! In Toronto, wealthiest city in the second wealthiest country, ambulances are careening from hospital to hospital in a desperate effort to find an emergency bed, and cancer and cardiac patients are forced to wait for treatment for unacceptably long times.

Recent immigrants will be in no position to take care of the old members of their families out of charity, even if they were so inclined. And what are the old to expect of their grandchildren-if any—increasingly raised in single-parent homes, or, in the case of a richer minority, by nannies from other cultures?

A materialistic society increasingly spiritually empty and disappointed by the Golden Goose's inability to increase the golden egg, a society not reproducing itself and so having to turn over its institutions to immigrants who understand only superficially the deeper sense of its traditions, may not be in the best frame of mind for dealing with the problems that are building up. Will the citizens allow their democratic governments to make the painful adjustments necessary to avoid the semicollapse kind of rebalancings like the 50 percent devaluation of the Mexican peso, which can lead into depression and calls for *simpliste* solutions of an authoritarian sort?

Such a dreary scenario is in no way an inevitable result of the HTX. Indeed, as I said, the consciousness-raising capabilities and the wealth generated by the HTX give solid hope of finding our way out of such a sad debacle. The efficiency and "having rather than being" *tendencies* in the being of the HTX do not exclude the possibility of our individually and collectively coming to our senses. Good sign: Already many governments are addressing their deficit problems much more seriously than they were even ten years ago, but an era of profligacy has left for future citizens a staggering accumulated debt and high taxes. That narrows the room to maneuver come the next recession.

20. Quoted by James Dale Davidson in "Strategic Investment" newsletter, June 21, 1995, p. 11.

Dreams of Revival of Sound Spirituality:
Back to Eisenhower and Norman Rockwell?

Dream for a moment of what a massive revival of sound spirituality could do for the wealthiest HTX societies. "Soundness" is of course in the eye of the beholder, so let me spell out what I would find "sound": I mean a spirituality that would have as an effect to undergird greater commitment to family life, with attendant seriousness in educational reform; a desire to save more for the future of oneself and the children, hence a more sober lifestyle (which providing savings results too in a lower cost of capital for the society, which just happens to contribute to creating more jobs, which creates a more moral atmosphere than unemployment—there is something to be said for "the Protestant work ethic"); a greater willingness on the part of families out of love and respect to take care of their own, especially the elderly, instead of dumping them on hugely inefficient and uncompassionate state-financed institutions (hence contributing to lowering government deficits, which again lowers capital costs); a diminution in private and public mediocrity and dishonesty of the kind that selfish materialistic lifestyles, driven by greed, encourage (and that inner struggle against mediocrity would increase efficiency, lowering costs of goods and services).[21] Less wasteful consumption, fewer restaurant meals and less tourism, as grandparents stayed around to enjoy their grandchildren, would further increase savings, lowering the cost of housing and making more money available for higher quality education. More wives having several children and staying with them, at least in their early years, but also being there when the teenagers come home from school, would reduce unemployment and delinquency, and would provide a richer, more personal education for the children. Less television, which would diminish if the mothers have the support to be able to spend much time on the children's education instead of being pressed by events to use the boob tube as babysitter, would mean less exposure to systematically encouraged coarseness and violence and vapid mindlessness. Parents who are present can encourage children to play, which increases their inventiveness. (Expert voices are decrying too much use of computer games.) The parents striving to provide a more relaxed home environment would create a place where many would enjoy eating, discussing, and playing, reducing the money and time wasted on

21. Both mediocrity and dishonesty saddle the society with waste, underperformance, and poor allocation of resources, including the huge costs of pursuing, convicting, and interning criminals.

going to movies, watching mindless television, and escaping to expensive, mediocre restaurants. Families getting together more might free up escape time that could be used instead for the all-important volunteer activities that spread love throughout the community. Strong church communities and more involvement in the schools would enlarge the circle of friends and deepen seriousness about life. And religious *engagement* should increase the sense of the need for active concern for the suffering in the community, as well as for fearless honesty and sobriety.

Ah, the Eisenhower era, old-fashioned and very chauvinistic, a retrogression to the bourgeois start of the industrial revolution. Remember, in the HTX, we do not have anything to learn from the past.

Dear reader, you would not have stayed with me to here if you thought I was incapable of realizing that the Eisenhower era's own mindlessness helped bring us the mindlessness of the full-blown HTX phase two. I must admit, however, that I believe the real Okie country-western we snuck under the covers to hear on AM in Tulsa was culturally less degenerate—I'll stand "Johnny Lee Wills and His Texas Cowboys" against the "Stones" any day. The first was wonderfully mindless, the second not mindless at all—deliberately nihilistic, *fun* versus *hate*.

Of course, I am not so naive as to expect such a retrograde vision to go down well with many HTXers. But if careful reflection shows such a family lifestyle still to be the healthiest—including the health of the economy, and entire books have been filled with the sociological, economical, psychological data, without which no one in the HTX is prepared to accept anything—then those who share my concerns be consoled! The HTX, while not fostering such a spirituality, *does not inherently preclude it either.*

For any of this reinforcement of the human foundations of the HTX to work, managers would have to make one objective of their planning helping to keep families intact and rooted. Totally drained fathers and mothers cannot contribute their full share to the family. If managers will reflect on what floating employees and "associates" depressed by family breakdown cost the enterprise (not to speak of the state), then they might be prepared to make more farsighted concessions.

(Since I wrote this, one of our daughters, who is a general manager in a company with 24,000 employees in Canada, was urged by her superiors to allow them to put forward her candidature to head a new nationwide department, directly under the vice president, to group all dealings with the environment [in which she has managerial experience], health, safety, and a new section "to assist employees to balance work and family responsibilities." Reflecting on her own family responsibilities, with two small boys at home and a schedule she arranged allowing her to be home after school, she did not apply. In the family discussion leading to the decision, she wryly commented that it would be "pretty difficult, from a staff position, to convince harried managers rushing to meet deadlines and contain costs to divert much attention to solving employees' personal conflicts." I agreed. In optimistically insisting that the HTX is not inevitably dehumanizing, I never suggested that finding and making prevail humanly sound solutions in concrete situations would ever be simple.)

In the last pages I have concentrated on the personal-familial base of human existence. But throughout this study we have seen how our involvements in the other worlds, those of work, of the society, of the nation, of civilization, and the pressure from all kinds of invasive HTX forms can both erode much of what we most dearly desire on that personal, familial, and religious level, and, worse yet, at times crash into it like a huge, shocking pee-ball meteorite, as happens, for instance, when I suddenly lose my job or when terrible dislocations, perhaps even political unrest, occur in the society.

Some "aesthesis of compartmentalization" is unavoidable. When one is involved at the family level, for instance spending an intense Sunday afternoon discussing marital tensions with a daughter and her husband, it is interesting how little one really cares about how one can call up on the Internet color photos of the newly discovered Stone Age cave in France, or maybe even canned marital advice. One cares not a hoot about the trials and travails of just-in-time inventory; and one takes the phone off the hook. Of course HTX considerations come up in the discussion (e.g., the temptation to flop in front of the TV during those few minutes when, the children at last in bed, the parents could have a quiet moment together). And I did remark to those young persons that it is very much HTX management style to structure one's day willfully to provide that quiet half hour of relaxation. But I reminded them that, living as we do in a highly HTXed society, you have to take advantage of its means to provide, somewhat artificially, those old-fashioned human times. I know businessmen who write into their daily calendar a half hour for prayer and meditation, a time scheduled for God! Even the Almighty cannot reach him electrically: the phone is disconnected.

Developing a Strategy

Somewhat bad-humoredly I brought up early on in this reflection the need for management strategies for our lives. Is it not also incumbent on those who can divine a bit of what is coming upon us politically, economically, and socially to develop an investment strategy, an educational strategy for the children, and to work strategically to find some serious base for community before the social structure is undone by senseless violence fed by nihilism?

But there is a note of optimism in the premises here. Thinking in such planning terms implies that the being of the epoch is to some degree malleable. Does the ontology that has begun to emerge from this preliminary reflection support the belief that, no, we are not fated to be swept wherever being will go, but rather, yes, to the extent that being's illuminations enable us to see certain trends and to understand possibilities, we enjoy freedom at least to try to clarify the vision of where, within the limits of the possible, we would like things to go and to attempt to persuade others to cooperate to move the situation in the desired direction?

I believe the ontological foundation of our freedom includes being's revelation of the openness (limited, to be sure) of the situation, and of the need and possibility to gain perspective on it. That perspective includes (*pace* my friend designing the virtual corporation) historical depth, which enables us to see trends rooted in the past and to understand better the existing structures.

Such a search to understand the situation entails assessing the way well-entrenched cultures interact with the civilizations imposed on them and molding them, and molding me as one who lives through those cultures. It requires attending to how both cultural and civilizational depth transform the HTX forms that come in upon them "from on high." More primitive and recalcitrant than all those cultures, there remains human nature. I believe it wise, in the present introduction to the HTX, to give human nature the last word.

Conclusion: As We Are Swept along by the HTX, How Free Are We?

An optimistic reader of this manuscript—a high-tech manager—offered the opinion that, given that human nature has not changed much, and that the HTX rather opens a greater wealth of possibilities, the individual is now offered more opportunity to exercise his freedom than ever, and this more than offsets pressures in dangerous directions, which he admits are serious. Let us rally around!

I wish I could share his optimism. Yes, basic human nature exercises its molding power throughout the great upside-down HTX superstructure, just as it has when tribal structures, then civilizations reigned supreme. All the discussions whizzing over the Internet from Singapore to Vienna are utterances and opinions of free individual persons, many showing wonderful imagination and creativity. And it remains as true today as it was for Abel, you cannot be forced to do anything, you can always "opt out," if only, *in extremis*, by sacrificing (lit. "making holy") your life. Even psychotic compulsion can be treated, if there is a will to address it.

But man always ends up having trouble seeing the world any other way than the local socially prevailing way. One's "mind-sets" and motivational attitudes may change, but some prove pretty persistent throughout a lifetime, and necessarily mesh well with the milieux in which one either chooses to immerse himself or simply cannot much avoid. For all and sundry, *I am* many of those ideas and pursuits incarnate, they constitute to some considerable measure who I am, and to some extent my body (and my clothes) express what they have made of me. So maybe I am not as "free" as that great menu of HTX-generated possibilities makes it seem.

Because we have to adjust to the new HTX challenges or die (and hundreds of millions who did not succeed in recent decades, for instance finding themselves in the hands of Stalin's MKVD or in the way of Hitler's armies or Mao's program of *debourgeoisification*, have been killed in the process), those of us still alive have not only learned new languages and skills, but we have also had to get good at putting the paralyzingly frightening implications of some of these new inventions out of our minds—or, better, sanitizing these realities by accommodating them psychologically in some way.

I shall illustrate this by indulging in a patently cheap shot.

I was watching the film *Crimson Tide,* and it occurred to me that, yes, it is true, one Trident submarine can launch from below the sea's surface missiles with explosive power exceeding the accumulated power employed in all the wars since the gunpowder revolution. And yes, a very limited commander with the accord of his executive officer could launch those missiles, with results totally beyond our imagination, and of a sort we would prefer not even to think about. How can we live with these realities, with the thought of 20,000 Russian warheads, any one of which could get sold to an evil source by some greedy or desperate commander in one of the fragments of the Soviet army? Behind this horror lies the reality of a newness that results from a quantum leap in capability: For millennia we could raze whole villages in minutes, then since fifty years bombard whole cities in hours, but now arguably eliminate mankind itself through a nuclear winter in an hour of missile exchange. The scale of power concentration that we have attained is beyond anything we can relate to emotionally.

We all know about the possibility of "nuclear winter," or the possibility of "nuclear blackmail" (and any honest person most admit the threat has not disappeared with "the end of the Cold War.") When someone is *gauche* enough to bring it up, very reluctantly we acknowledge the reality of the threat, and then clang our minds shut. Recall the longtime friend of mine who spent the last third of his working life worrying himself silly about it, but then he was number three in the Arms Control Agency of the U.S. presidency, the official who commissioned the two major studies on the topic, one by the Rand Corporation and the other by Strategic Initiatives, Inc., the first concluding that mankind was not likely to survive all-out nuclear war and the second that an important remnant would likely survive, but in the southern hemisphere only.

No one will deny the importance of somehow trying to control the genie, even if putting it back in the bottle is impossible. But we know that we personally cannot do much about it (boycotting French wines after a bomb test in the Pacific is one of those gestures that makes one feel righteous but basically does not address the problem), so we hope competent people in our governments are managing to attract the harried decision makers' attention at the right time to get them to take steps to avoid "nuclear proliferation." With that, we just leave a little festering disquiet in the depth of our souls, where it likely will lie until the day the news bulletin announces some crazies have said they smuggled a nuclear device into Manhattan and are demanding x, y, and z. (It is amusing that the equivalent of that for the terror of designer genes was the innocent sheep, Dolly, named because of her creator's "admiration for Dolly Parton's mammary glands." [The DNA for Dolly was extracted from the breasts of the sheep being cloned.])

The apocalyptic gospel passage that ends the liturgical year shows Jesus warning the disciples about the end of the world and telling them no one but the Father knows the day and hour, but be prepared every day. With our managerial mind-sets we are constantly "programming" processes in view of more or less dimly perceived goals, and at the same time having to shut out of mind an increasing catalogue of ever clearer menaces ("the debt overhang," "the population explosion," "the population decline of the First World," "the break-up of Canada," "the now sure ravages of the AIDS epidemic," "the Asian flu," "the disintegration of Russia," and all the coming bad fruits of the worsening social indicators mentioned in chapter 1.)

Human beings have always had to shove thoughts of catastrophic possibilities (or, like death, inevitabilities) to the margins, mythologizing some of them as a way of coping. What is new in our situation in this regard? First, so much more is foreseeable. On top of that, we have invented horrible new possibilities of such devastating potential alongside the wonderful ones. And being our own creations, they are somehow very difficult to mythologize away (*Dr. Strangelove* did not make M.A.D. go away.) This all leads, in the more responsible, reflective souls, to the gnawing sense of bad faith I mentioned earlier: there is so much we as individuals should be addressing better that we cannot, or to which we just refuse to give due attention. The activist sense leaves us with the impression we should really be doing more. Plunging into our work is a good way of pushing neglected family, spiritual, political, and social works to the margins. For the less responsible, especially grandparents who have time to help, there is tourism.

But why exactly do the old means of placating the gods and constructing mythological explanations not work anymore with our expensively purchased disastrous inventions? To start with, Christianity, the dominant old tradition of the West, from its Hebrew origins, is not into the mythologizing business, it is resolutely antimyth. (Heretical cults often deviate in that regard.) Far from being about escape, it places personal responsibility at the center, true responsibility in the sense of love. The Creator replaces Fate, a *Divina commedia* replaces Greek tragedy, but the way to salvation being realistic is beyond us and calls for the strongest medicine, because overcoming the ravages of our bad free decisions is a work of divine proportions.[22] As always in Christianity, we are confronted at the center with paradox: "Pick up your cross and follow me!" has to be played off against "My yoke is light." These are hard to balance. When Christians deform the essential message by tailoring it to their personalities, then either the Good Shepherd is allowed to obscure entirely the Crucified, or, if one has a sour disposition, a triumphant Puritan asceticism coupled with a fanatic work ethic obscures the love of a Fatherly providence. The way of true loves requires that the old man die, and dying to self is a lifelong hard struggle to "put off the old man," struggling against pride, the temptation to think we are God, as one learns to be of service to others, abandoning all destructive selfishness, and with it the last vestiges of the illusion that we can create life,

---

22. *The Catholic Tradition* offers throughout evidence for the claim of antimythology. It also criticizes Catholic ideologies that would interpret the teaching authority of the Church in a way that diminishes freedom. This is precisely *not* what the Church teaches.

save the world (by undemonizing our bad constructions), or *earn* everlasting existence. That vision entails then doing away with every last one of the HTX temptations to play God.

The HTX man is a sitting duck for sects, by which I mean *unsound* spirituality, because most sects are do-it-yourself religions invented by strong leaders who demand submission, tyrannical fathers who relieve of responsibility tired individualists who have become confused about where they are going. Typically people who enter sects get cut off from their families, the family does not need them, and they do not need the family. The real God is not like that. He calls us, not to escapism, but to maturity, to full self-responsibility, which can only happen in realistic circumstances: the real self who realizes the responsibilities incumbent upon him by virtue of his situation, the need to serve others because their needs are pressing and real. It is not about escaping from the world, or giving over decisions to others, it is about ceasing to lie to oneself in order to embrace the hard reality presented to us immediately by those who have a legitimate call on our energies. Recall that response-ability, from the Latin root, *spondeo,* means "I commit."

The ability to respond to the challenges of the present situation entails the commitment to build families and community, families not closed in on themselves and communities, like parishes, that are welcoming to all, a countertrend to our animal tendency to flock in order to exclude. The Church is serious when it proclaims a genuine catholicism—all truth is to be shared with and by all people, the gathering of one family of man built upon the embrace of all the truth science can discover, the traditions have preserved, and the arts glorify. Anything short of that would be an elaborate idol, not the in-pouring of the Infinite Source of all creation. The new wine bursts all old wineskins.

How Might Genuine Religion Transform the HTX?

What kind of effect could such a turn to genuine religion—not idol building—possibly have on the massive power structures of the HTX, and through inflecting them, on its being? Reflect on those power structures for a moment, and consider how they interpenetrate.

At the base the individual family remains the single most powerful formative influence on the future through the depth of formation it gives. But one family reaches into the souls of only a few individuals in a generation, although the enduring effects of the children operating in their respective societies and passing on their strengths (and weaknesses) over many generations can have a powerful multiplier effect.

Healthy families grow exponentially, to the horror of people with the "Gospel of Death" mentality of nihilism. (While the world does not "need" more people, it can always use more good people, the kind who devote their abilities to making it a better place for us all.)

In the great hierarchies of the national governments and the massive corporations there is awesome leverage to be found. At this moment we are witnessing in Ontario the power of a resolute premier to change the course of the economy and the vast social welfare behemoth of Canada's wealthiest province. Whatever one may think of his ideology and policies, it is daunting to see firsthand what one man can do when he has a clear vision of what he wants to accomplish and the institutions of government allow him to shake the structures to their core. Similarly, it is impressive to see the power wielded by a Lou Gerstner, backed by the board, in turning inside out what everyone thought to be the hopelessly inflexible behemoth of IBM. I am not passing judgment on either—I am just calling attention to what one resolute person can initiate.

In the Church, strong bishops are able to accomplish much, but they are nothing without a prayerful laity, and they are the first to admit it. There the true "leverage" lies in the depth of sanctity of the individual soul.

In a philosophy of hope, respect and awe are fostered for the mysterious depths of being and nature, generating an attitude of listening (*obedire = audire ad* = obedience, not an notion treasured by HTXers, who think they are to create everything *de novo*), listening to the authentic and liberating voice of every form of genuine revelation, from those yielded up by subatomic particles provoked by accelerators to those of Transcendent vision. These ultimate illuminations, which move us to be receptive and appreciative, are a gift, a grace—fruits of human devotion and the gift of the things themselves, from particles and genes, through symphonies and sacred texts, and from illuminations granted deep within the human soul.

Appreciation for what is given is that ultimate theme about surviving in an age of virtual reality which in the present brief work has been orchestated with but a few meager examples. Meanwhile, I have been at work trying to make a more substantial contribution to showing how, while living amidst the HTX, earnest efforts at "appropriation" can offer enhanced appreciation. As promised in *Tradition and Authenticity in Search of Ecumenic Wisdom,* I have published a critical appreciation of a tradition that has formed me, and has been perhaps the single most important formative factor in Western civilization, still very much alive in these HTX times, *The Catholic Tradition*. I also promised to explore the rather different problem of achieving profound appreciation, with comparison to my own, of a great tradition foreign to me. I am finding it daunting indeed to get a halfway adequate hold on Islam, but the effort is proving exhilarating and profitable to my own search for wisdom. Finally, throughout this series, I have brought out, as you see here, the centrality of the anthropological dimension, the fundamentality of man, so now I have taken many of the anthropological themes that have woven through *Being and Truth* and *The Catholic Tradition* and developed them into an extensive treatise that has gone through several drafts. The very tentative nature of some of the anthropological and epistemological considerations raised in this present introduction to the question of the essence of our epoch are being explored more sustainedly and in an order suitable to a less inadequate illumining of human being.

Index